KT-509-572

BLACKWELL GEOGRAPHY PROJECT 2

The Developed World: Europe, North America and Japan

© 1988 Neil Punnett, Peter Webber, Stephen Murray

First published 1988

Published by
Basil Blackwell Ltd
108 Cowley Road
Oxford OX4 1JF
England

All rights reserved. No part of this publication may be
reproduced, stored in a retrieval system, or transmitted in any
form or by any means, electronic, mechanical, photocopying,
recording or otherwise, without the prior permission of Basil
Blackwell Limited

British Library Cataloguing in Publication Data

Punnett, Neil
    The developed world: Europe, North America
    and Japan. – (Blackwell geography project; 2).
    1.   Great Britain. Geography.
    I. Title   II. Webber, Peter   III. Murray, Stephen, *1948 –*
    910′.091722

ISBN 0–631–90094–2

Designed by David Chaundy
Typeset in Melior and Helvetica
by Opus, Oxford
Printed in Hong Kong by Wing King Tong Co. Ltd

Acknowledgements

The authors and publisher would like to thank the following for
permission to reproduce photographs in this book.

Ackroyd Photography 2.7B; Laura Ashley 4.6B; Assiport 4.4B;
Associated Press 3.5A; Barnaby's Picture Library 2.1B (pine trees),
2.5A and B, 2.6E, 2.9A, 2.12B, 3.2A, 3.9C, 3.10C, 4.7E, 4.15A; A
Bishop 4.13A and B; D Bishop 2.4B, 4.14A, 4.16C; British
Aerospace 4.5A and B; British Museum 2.30; Camera Press 4.7B,
4.9C; Mike Corby 2.11B; Cumbria Tourist Board 3.2B; P J Downes
1.11A; Dutch Ministry of Agriculture and Fisheries 4.2B; Eurotun-
nel 1.5B, C and D; Ken Foulds 3.10E; French Government Tourist
Office 1.10A (Pompidou Centre, Sacre-Coeur Basilica and Notre-
Dame Cathedral); Keith Gibson 1.3C; Ambrose Greenway/Freight
Forwarding 1.13B; InterTAN UK Ltd. 4.15C; Italsider Steelworks
1.12B; Japan Information Centre Introduction, 4.14C, 4.16B; Eric
Kay 1.2C (forest and sandstone mounts), 2.1B (x4); Komatsu UK
Ltd. 4.14D, 4.15D; London Transport Museum 1.2C (London
traffic); Lufthansa Photo Archive, Cologne 1.13A; NASA 1.13D;
Panos Pictures 2.12A, P&O Ferries 1.4B (ferry); Neil Punnett 1.11C,
3.9B; Rockware Reclamation 3.1A; Sollac Dunkirk 3.6C; Spanish
Embassy 4.3C; Spectrum Colour Library 1.13C, 2.9D, 4.7D; Richard
Spilsbury 1.10A (Place de la Concorde); Statoil 3.4C; Louise
Stewart 1.10A (Eifel Tower), 1.11B; Swiss National Tourist Office
3.3C; Usinor/Sacilor 3.6B; Venice Tourist Office 2.10A, B and D;
Peter Webber 1.6C, 1.7B and D, 1.8B, D and E, 4.8; Westerly Yachts
Ltd. 1.11D.

BLACKWELL GEOGRAPHY PROJECT 2

The Developed World: Europe, North America and Japan

Neil Punnett, Wilberforce College, Hull
Peter Webber, Kingsdown School, Swindon
Stephen Murray, St David's School, Wrexham

Basil Blackwell

Contents

PEOPLE'S LIVES AND WORK

Pupil's introduction

Geography is a vital, living subject. Geographers study the world and its people. They look at the environmental, physical and social conditions that affect how we all live. The study of geography helps us understand important issues about people and the environment.

This book covers some of the major issues facing our world. It focuses on the **developed world** – Europe (including the USSR), North America and Japan. If you read the newspapers and watch TV news broadcasts, you will see that the developed countries dominate the headlines. This book looks at some of the reasons why these countries have grown to be so rich and powerful. The USA, in particular, has a world-wide influence. When was the last time you drank a can of Coca-Cola, ate a Big Mac, listened to an American pop group or watched an American TV film?

Look at the *Contents* list on page 4. The four main topic areas are about our everyday lives. **Settlement** means where people live. **Environment** refers to the surrounding area, eg mountains, the coast, the weather. We all have many **needs** – for shelter, food, water, fuel, power, transport. . . These needs affect our **lives and work**. Most developed countries are highly industrialised. We are surrounded by science and technology at work and in our leisure time. Most people in the developed world have enough to eat, but many are unemployed. . . These are just some of the issues this book asks you to think about.

Look at the photograph on the opposite page. A geographer can understand a lot about a place by looking at a photograph like this, especially if it is linked with a map of the area, showing street names, directions and so on. You can probably write something about the photograph already. Try answering these questions:

● Where in the world do you think this place is?
● What type of settlement is it?
● What is the environment like?
● What do the people need?
● What type of work might the people who live here do?
● What do you think it is like to live here?

To answer these questions you will need to learn the **key words** that a geographer uses. You will also have to practise geographical **skills** like map reading, map drawing, field sketching and graph drawing. There will be new **ideas** or **concepts** for you to understand. Sometimes the subjects you study will not have a right or wrong answer. You will have your own views, but the geographical **issues** will need to be discussed.

The box in the top right-hand corner of each unit in the book lists the skills, key ideas and issues dealt with in the unit. The matrix charts at the back of the book will show you how these fit together and develop as you work through the units. They also tell you where you can find the same subjects, issues and concepts elsewhere in the book.

When you have found out about the developed world, you may want to know more about life in developing countries. Book 3 in the series deals with the developing world.

Geography is important for all of us. It is also fun! So make the most of your opportunity to learn more about the world and its people, so that you can understand the opportunities and problems that face us all.

1.1 Europe in maps

An **atlas** is a book of maps which is a very useful source of information for geographers. Atlas maps contain many different types of information.

The land masses of the world are divided into seven **continents** (see **A**). **Europe** is one of the continents (**B**).

---

1 Look at **B**. Find a **political** map of Europe in your atlas (look in the Contents list) and use it to help you fill in a table like the one started here.

Country	Name of country
1	
2	
3	
....	

---

2 What countries of Europe are:
a in the EC
b East European states
c in neither the EC nor Eastern Europe?

3 Use the outline map of Europe in the *Activity pack* to make your own political map of Europe. Fill in the names of the countries on your map.

4 What information does each map (**A**, **B**, **C**) give you about Europe?

The continent of Europe stretches from the Atlantic Ocean in the west to the Ural Mountains in the east, and from the Mediterranean Sea in the south to the Arctic Ocean in the north.

C is a **physical map** of Europe showing the **relief** (height) and **drainage** (rivers) of the continent. What colours are used to show high and low ground on **C**?

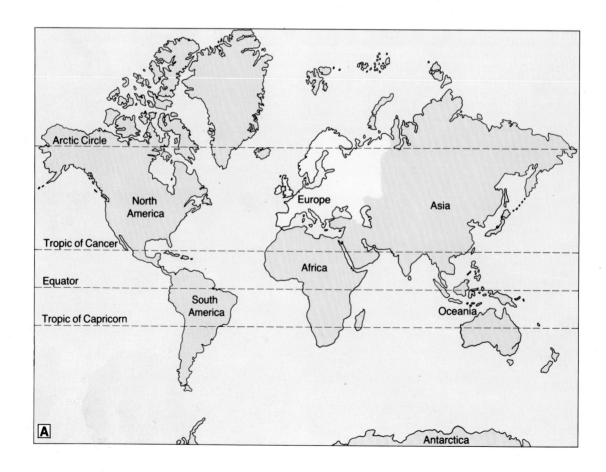

Skills	using atlas, mapwork
Concepts	Europe, continent, political/physical map
Issues	regions, EC, political divisions

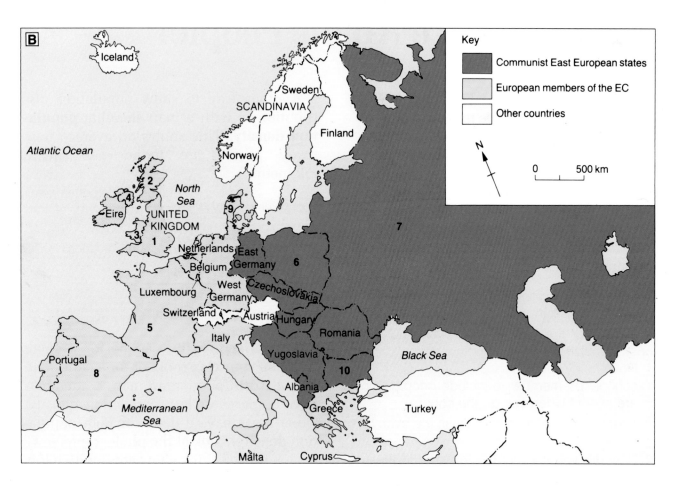

B

Key

■ Communist East European states

■ European members of the EC

□ Other countries

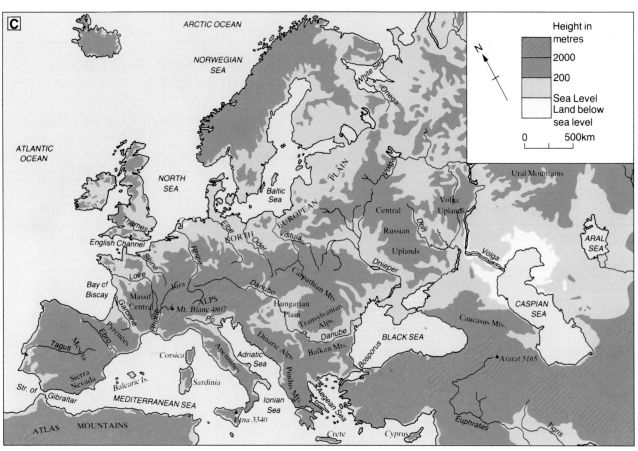

C

Height in metres

2000

200

Sea Level

Land below sea level

1.2 Where are All the People?

How many people are in your school? If you try to find out you will be making a **census**. A census is carried out in Britain every ten years (1981, 1991 etc) to find out how many people there are in Britain, and where they live. The census asks other questions too, such as: What job(s) do you do? Do you own a car? Do you have an inside toilet? The census builds a picture of the **population** of Britain.

Read the questions first and then discuss with your teacher the best way of finding out the answers.

1 How many people are in your class in school? (Remember to include people who are absent when you do the count.)
2 How many people in your class are over 13 years old?
3 How did you find out the answers to these questions?
4 What problems did you have in getting the information?

Geographers are very interested in how many people there are in the world and where they live. **A** is a table of **world population distribution**. The information was gathered by using the census of every country in the world.

Another way to show population distribution is with a map showing **population density**. This shows on average how many people are found per square kilometre.

$$\text{Population density} = \frac{\text{Total population of an area}}{\text{Total area in km}^2}$$

For example the population density of Africa

$$= \frac{560 \text{ million people}}{30.2 \text{ million km}^2} =$$

18.5 people per square kilometre.

Population density can be for a small area like your classroom or for a continent. **B** is a world population map. Some areas have a **sparse** or low population density. Other areas have a **dense** or high population density. Look at the photographs in **C**. Which do you think show areas with a low population density? Which show areas with a high population density?

---

5 Find the population density for the continents in **A**. (You can use a calculator to help you.)
6 What is population density and how is it calculated?
7 What does 'uninhabited' mean?

A World Population Distribution and Land Areas

Region (continent)	Population (millions)	Land area (million km²)	Population density (People per km²)
Africa	560	30.2	
Antarctica	Uninhabited	14.2	
Asia	2935	4.4	
Europe (including USSR)	780	11.4	
North America	265	24.3	
Oceania	25	8.5	
South America	436	17.8	

(NB Oceania includes Australia, New Guinea, New Zealand etc)

Skills	*survey, interpreting tables, calculating population density*
Concepts	*population distribution/density, census*
Issues	*hostile/friendly environments*

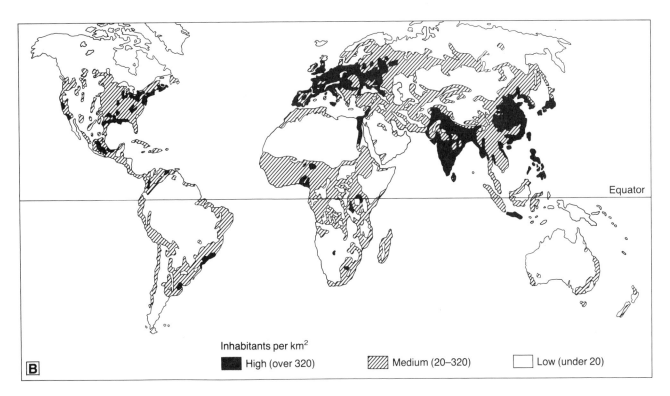

Inhabitants per km²

■ High (over 320)   ▨ Medium (20–320)   ☐ Low (under 20)

B

C

8 a Study map **B** and the photographs in **C**. For each photograph, say whether you think the area would have a low or high population density, and why.

b What other reason can you think of for why an area may have a low or high population density? (Think about what makes an area hostile or friendly to people.)

1.3 Europe's Links – Migration

Have you any relatives living overseas? If so, where do they live? A group of pupils did a survey to find out which countries their relatives lived in, and why they lived there. **A** shows the results of their survey.

Country	Reasons
India	parents' families live in India
Cyprus	father's family is Cypriot and lives in Cyprus
Kuwait	uncle is an engineer there
Australia	uncle and aunt moved there
Canada	aunt married a Canadian
USA	eldest brother is working there
India	grandparents are Indians and have always lived in India
Hong Kong	uncle teaches in a British school there
Poland	great grandmother is Polish and lives in Warsaw
West Germany	brother is stationed with British army
Kenya	sister is a Voluntary Service Overseas (VSO) worker in Kenya

A

---

1 a In a small group list the countries where your relatives live.

　b Which ones are European countries?

　c Write down any special reasons why the relatives live in those countries.

The word used for people moving to live in a different area is **migration**. People migrate from one part of a country to another but when they move from one country to another they **emigrate**. When the pupil's aunt and uncle moved to Australia they emigrated *from* the United Kingdom. When they arrived in Australia they were known as **immigrants** as they had entered *into* the country.

Every country in Europe has experienced migration. **B** is a **flow-line map**. It shows where immigrants into France have come from. The width of the arrow represents the number of immigrants.

---

2 a What is an immigrant?

　b What is an emigrant?

3 Where have most of France's immigrants come from (measure the width of the arrows)?

4 Approximately how many immigrants have come from **a** Spain **b** Tunisia?

Why move countries?

Many immigrants into France, like the people in **C**, have come from poorer countries. Why have they moved to France? Think of France as a magnet which attracts people towards it. The reasons for the attraction are called **PULL factors**. The reasons for leaving a country are called **PUSH factors**.

---

5 Study **D**. Suggest reasons why many Algerians have moved to France. Write your reasons in two columns, like the ones below.

PUSH factors	PULL factors
1 Not enough hospitals	1 Good health services
2	2

Difficulties for immigrants

New immigrants in many countries face problems. Even after years in their new countries they suffer **disadvantage**.

Skills	groupwork, using flow-line map, expathy
Concepts	migration, 'push and pull, disadvantage
Issues	racial prejudice, disadvantage

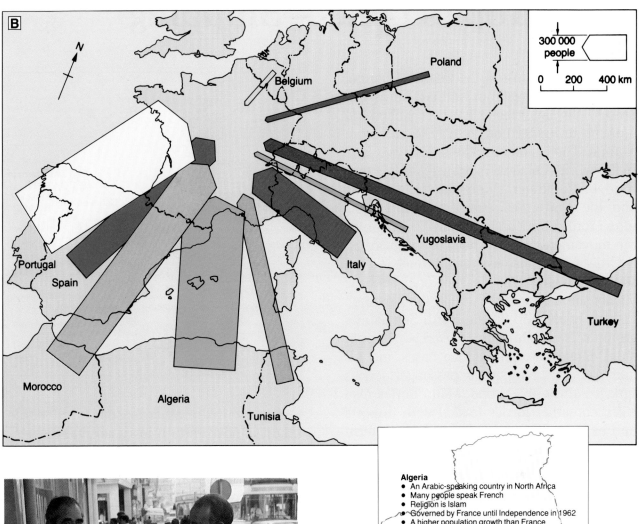

Algeria
- An Arabic-speaking country in North Africa
- Many people speak French
- Religion is Islam
- Governed by France until Independence in 1962
- A higher population growth than France
- Health services less developed than France
- Poorer farming than France
- Not much industry until recently
- Oil and gas exports
- Less education than in France
- The south is a hot dry desert

D

- Immigrants are not well received by everyone
- They suffer **racial prejudice** (eg Algerians do not get jobs because they are Arabs)
- They may have language difficulties (eg Algerians speak Arabic as their first language)
- They may have a different culture and religion (eg Algerians are Islamic)
- Immigrants can only afford the cheapest housing
- The police may pick on them

1.4 Europe's Links – Shipping

Have you ever crossed the English Channel by ferry? Shipping from all over the world passes through the Channel (called *la Manche* in French) on its way to and from Europe. The Channel is the busiest sea lane in the world. Oil tankers bring oil to Western Europe. Container ships use the Channel and the North Sea. Small coastal vessels move round the coasts of Europe. **A** shows the movements of **merchant** or cargo ships.

Other vessels also use the Channel and the southern North Sea. Pleasure craft such as yachts sail across and through the Channel. Fishing boats work in the North Sea fishing grounds. Naval ships go on exercise. Supply ships move between service bases and oil and gas rigs. Numerous passenger ferries link Britain with Europe. Many ferries also carry container lorries loaded with imports and exports. Dover is Britain's busiest ferry port, with passenger ferries to French and Belgian ports.

---

1 Make a keyword plan of the different types of shipping using the Channel and North Sea.

Choice of routes

There is a wide choice of ferry services from Britain to Europe and much competition for passengers among the ferry companies. People choose which service they want for several reasons:

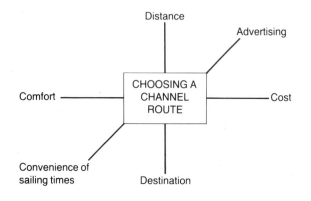

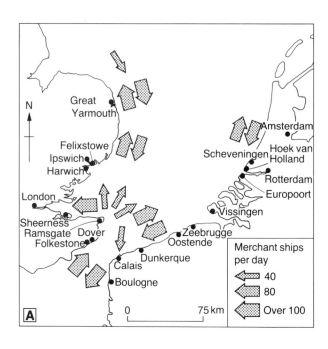

Can you think of other reasons why people might choose certain ferry services?

---

2 Which route to Europe do you think the following people should take? Base your decisions on the distance to the ferry port.

   a Helen Ross from Edinburgh, wanting to go to Bergen in Norway.

   b The Nash family going from London to Cherbourg.

   c Robin and Mary Peel, from Exeter, going to Cherbourg.

   d Maurice Zapp, travelling from Norwich to a conference in Oslo.

   e Enrico Panelli, going from Exeter to Hamburg, on business.

   f Winston Johnson, going from Liverpool to Rotterdam.

3 What other reasons might people have for choosing a particular cross-Channel route? Add these reasons to a copy of the keyword plan, *Choosing a cross-Channel route*.

4 Which ferry route would you take to travel from your home to
   a Paris
   b Amsterdam?

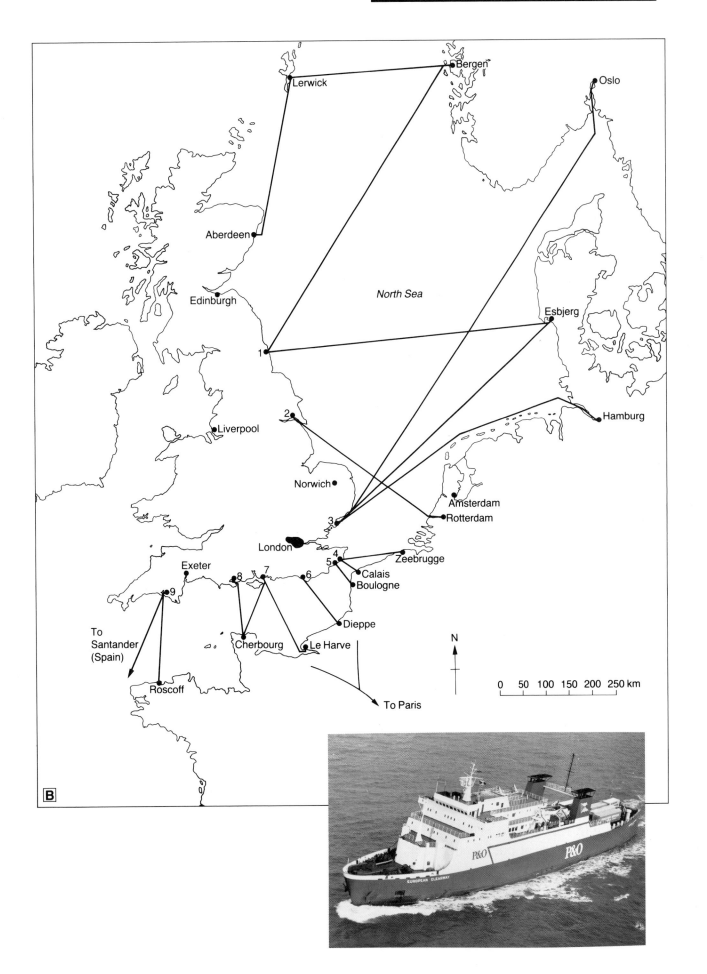

Lerwick

Bergen

Oslo

Aberdeen

Edinburgh

North Sea

Esbjerg

Liverpool

Hamburg

Norwich

Amsterdam
Rotterdam

London

Exeter

Zeebrugge

Calais
Boulogne

Dieppe

Cherbourg

Le Harve

To
Santander
(Spain)

Roscoff

N

0 50 100 150 200 250 km

To Paris

B

15

1.5 Europe's Links – The Channel Tunnel

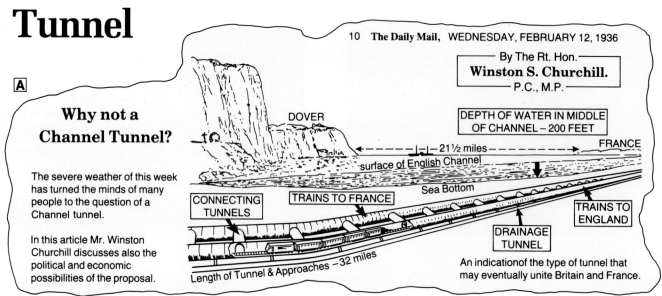

A

Why not a Channel Tunnel?

The severe weather of this week has turned the minds of many people to the question of a Channel tunnel.

In this article Mr. Winston Churchill discusses also the political and economic possibilities of the proposal.

10 **The Daily Mail,** WEDNESDAY, FEBRUARY 12, 1936

By The Rt. Hon. **Winston S. Churchill.** P.C., M.P.

DOVER

DEPTH OF WATER IN MIDDLE OF CHANNEL – 200 FEET

21½ miles — FRANCE

surface of English Channel

Sea Bottom

CONNECTING TUNNELS

TRAINS TO FRANCE

TRAINS TO ENGLAND

DRAINAGE TUNNEL

Length of Tunnel & Approaches – 32 miles

An indication of the type of tunnel that may eventually unite Britain and France.

People have suggested many ways of linking Britain with the rest of Europe: by sea, by air, by bridge or by tunnel. In 1936 Winston Churchill argued for a Channel Tunnel (**A**). In fact, as early as 1802 a proposal was put forward for a tunnel – which would have taken horse-drawn trolleys.

In 1986 a Channel Link Treaty was signed between the British and French governments. Construction began in 1987 and the tunnel is planned to open in 1993.

The joint British and French company building the tunnel is called **Eurotunnel**. There will be two single-track rail tunnels and a separate service tunnel (see **B** and **C**). They will carry through rail traffic and vehicle-shuttle trains (**D**). A drive-through link is also being considered, to open in the year 2000.

What are the advantages?

Supporters of the Channel Tunnel point to the following advantages:
- Trains will run at least every 20 minutes (every 5 minutes at peak frequency).
- The tunnels will take up to 400 vehicles per hour and over 40% of all cross-Channel passengers.
- The journey from London to Paris or from London to Brussels will take three hours.
- There will be better road and motorway links to serve the tunnels.
- The tunnel will help trade (60% of Britain's trade is with Europe).
- Building the tunnel will create jobs – up to 35 000 in Britain; more workers will be employed in constructing the shuttle trains.
- The tunnel will not have to be closed in bad weather.

---

1 What two countries will the Channel Tunnel link?

2 Why did Winston Churchill say, in 1936, that the minds of many were turning to the question of a Channel Tunnel (see **A**)?

3 How will people and vehicles be transported through the tunnel?

4 Use **B** to calculate the approximate length of the Channel Tunnel.

5 **a** How long will the journey from London to Brussels take?
b How much faster is this than the rail and ferry route (see Unit 1.4)?

6 Explain how the tunnel will
a help trade
b provide construction jobs.

Skills	groupwork, viewpoints, weighing evidence, decision making
Concepts	speed/efficiency, conflict of viewpoints,
Concepts	environmental pressure

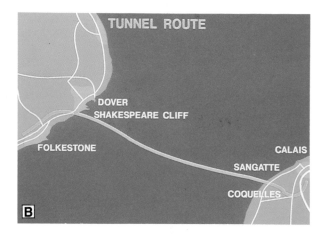

B

---

7 Why do you think roads in south east England will need to be improved when the tunnel is built?

8 Can you think of reasons why a rail tunnel was chosen instead of a road tunnel?

Not everyone is happy

Some people are against building a tunnel. The Nature Conservancy Council and the Royal Society for the Protection of Birds (RSPB) have protested at the suggestion that chalk waste from the tunnel might be taken to Dungeness and buried. These environmental groups are worried that the waste could upset freshwater supplies and damage bird habitats.

Some people are afraid that a tunnel will allow wild and stray animals from Europe to get into Britain. Some of these animals might be carrying **rabies**. This disease can be fatal for humans. There are strict quarantine controls at British ports and airports to stop infected animals coming into the country.

People living in the Kent countryside are also worried about the extra traffic that the tunnel will bring (leading to congestion, noise and pollution).

What other groups of people might be opposed to the Channel Tunnel?

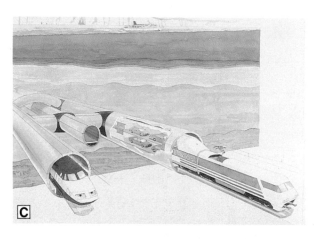

C

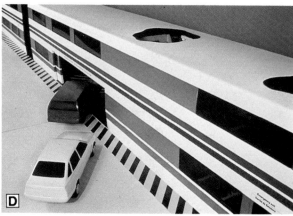

D

---

9 Why do you think the following people are against the Tunnel proposals:

  a The Channel ferry companies

  b The Channel ferry crews

  c People living in villages north of Folkestone

  d The farmer whose land is near the tunnel entrance in Calais?

 FURTHER WORK

In a small group, discuss the advantages and disadvantages of the Channel Tunnel.

● Write down your six main reasons for wanting a tunnel.

● Write down your six main reasons against having a tunnel.

● As a group, decide which are the strongest arguments for and against (either survey members' opinions or hold a vote).

● What did you decide? Write a short newspaper article arguing for or against the tunnel. Use your strongest reasons to support your argument.

1.6 A Visit to France

A party of 12 and 13 year olds went on a school trip to France. They travelled by P and O Ferries following the routes shown by the arrows on map **A**. **B** gives some information about times and costs of the journeys.

We often measure distance in kilometres (**linear distance**). We can also measure distance in terms of time (**time-distance**), or by cost (**cost-distance**). We use these types of distance in our everyday conversation. It is quite common to say, 'I live 10 minutes from school' or, that 'The town centre is a 20 pence journey from my home'.

---

1 Refer to the map and timetable (**A** and **B**) and use your atlas, to answer the following questions:
   a Name cities **a** and **b**.
   b Name the islands **c**, **d** and **e**.
   c What is the name of river **f**?

2 a How far did the school party travel from Portsmouth to Le Havre? Give your answer in kilometres.

   b How long did the journey take?

   c How much would the journey cost if you travelled as a foot passenger?

---

3 Now answer the same questions for the return journey from Cherbourg to Portsmouth.

4 Copy and complete these sentences. (Total the answers for questions 2 and 3.)
   a The linear distance travelled by ferry is .... **b** The time-distance is ....
   c The cost-distance is ....

5 What time does the return ferry from Cherbourg arrive in Portsmouth (remember the time difference)?

6 What route would you take from your home to the ferry port at Portsmouth (assume you are travelling by car)?

P&O European Ferries | **B**

Tariff for March		Timetable for March
	Single fares	
FOOT PASSENGERS	£	Departure
Adults	21.50	0900
Children		1100 — Day sailings 5½ hours
(4 and under 14)	11.50	1500 — Night sailings arrive
TRAVELLING WITH A VEHICLE		2200 — 0700 next day
		2300
Adults	20.00	**Cherbourg to Portsmouth**
Children		Departure
(4 and under 14)	11.00	1800 — 4¾ hours sailing time
VEHICLES		Time changes... Continental time is
4.00 m to 5.50 m	25.00	one hour in advance of the UK.

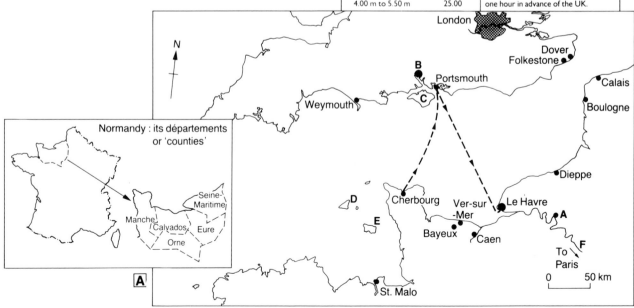

Normandy : its départements or 'counties'

Seine-Maritime
Manche Calvados Eure
Orne

London
Dover
Folkestone
Calais
Boulogne
B
Portsmouth
C
Weymouth
Dieppe
D Cherbourg Ver-sur-Mer Le Havre A
E Bayeux Caen To Paris F
St. Malo
0   50 km

N

A

Do you know anything about Normandy? When the school party left on the 0900 hours sailing from Portsmouth (see **A**) they knew very little about the **region** they were going to. They travelled to a field-study centre at Ver-sur-Mer. The aims of the course were to:

● map the route taken and note places of interest

● investigate the types of shops and shopping in the area

● discover how the people in the area earn their living

● observe the farming in the area around Ver-sur-Mer

● visit the Second World War landing beaches and museums

Whenever possible the pupils worked by **hypothesis testing**. (They began with a simple idea and then collected information and data to test if it was correct.) They made rough notes in the '**field**' (important for all good fieldwork). Back at the field-study centre the pupils wrote up the work neatly, using many different methods of presentation: descriptions, maps, field-sketches, graphs and tables.

What the pupils discovered

Normandy is one of the wetter and cooler regions of France. It is also near to Paris, which has a population of 9.3 million people who need large quantities of food.

Grass grows well and many dairy and beef cattle are kept. The black and white Friesian breed is common for milk production. The milk is sent to Paris or made into butter or cheese. The famous Camembert cheese comes from Normandy.

La Rosière farm in Ver-sur-Mer (**C**) specialises in beef rearing. It is 130 hectares in size and has about 80 animals. The farmer, Monsieur Calange, purchases calves which are bred on the surrounding dairy farms and feeds them up for beef. He keeps three breeds of cattle, the white Charollais, the red/brown Normandie and the brown and white Charollais/Normandie cross-breed.

The cross-breeds produce very good quality meat.

There are a lot of differences between Britain and France. France is more **rural** than Britain. More people work on farms and there are more villages.

C

---

7 Work together to design a questionnaire that you could take to La Rosière farm. Most of the questions should be general ones that you could ask any farmer in Britain or France, eg How many hectares is your farm? How many animals do you have?

8 Write a brief description of farming in Normandy.

FURTHER WORK

● Would you like to go to Normandy for a holiday? Give reasons for your answer. Why do you think Normandy has become a popular tourist area? (Use the evidence in this unit, and try to find more information in your school/local library. Local travel agents may have brochures or leaflets on France that include Normandy.)

1.7 Shopping in France

The pupils who went to France (see Unit 1.6) also studied shopping. They looked for **similarities** and **differences** between shopping in France and shopping in Britain. Although Ver-sur-Mer is a small village there are several shops. Certainly there are more than in a British village of a similar size.

Map **A** was drawn by a pupil. It was drawn without the help of an outline. All that was needed was a clipboard, paper and pencil.

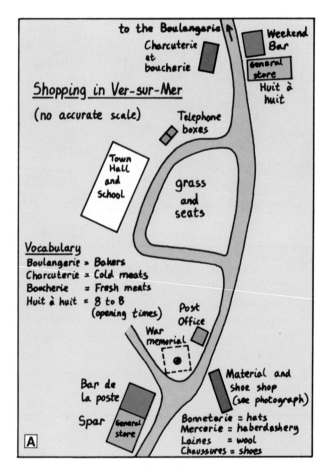

to the Boulangerie
Charcuterie et boucherie

Weekend Bar

General store

Huit à huit

Shopping in Ver-sur-Mer

(no accurate scale)

Telephone boxes

Town Hall and School

grass and seats

Vocabulary
Boulangerie = Bakers
Charcuterie = Cold meats
Boucherie = Fresh meats
Huit à huit = 8 to 8 (opening times)

Post Office

War memorial

Bar de la poste

Spar

General store

Material and shoe shop (see photograph)

Bonneterie = hats
Mercerie = haberdashery
Laines = wool
Chaussures = shoes

A

The pupils stayed at the **field-study centre** in Ver-sur-Mer. They noticed that the local people went shopping more frequently than they would do in Britain. People shop for bread (from the boulangerie) in the morning and afternoon. French bread does not keep well, so it is bought 'fresh'. In Ver-sur-Mer there are more unpacked fresh fruit and vegetables available than in a British village.

The Spar shop and the Huit à Huit are **self-service**. The other shops have people serving behind counters.

---

1 Study **A**. How many shops are in Ver-sur-Mer? (Include the bars and the post office.)
2 Which shops would not usually be found in a present-day British village?
3 What differences did the pupils notice between shopping in France and in Britain?

The pupils also mapped shops in part of the nearby town of Bayeux (**B**). The main street is different from most British High Streets. There are more smaller shops which **specialise** in one kind of goods. Some of the shops are for **short-term** or **daily** shopping. They sell bread, meat, cold meats and greengrocery. Other shops are **long-term** and people visit them less often.

B

4 Group these shops into short-term and long-term:

1	2	3	4	5	6	7
éléctroménagers	**disques**	**la boulangerie**	**la boucherie**	**coiffeur**	**marchand des fruits**	**la banque**
(electrical)	(records)	(baker)	(butcher)	(hairdresser)	(fruit shop)	(bank)

Present your results in the form of a table like the one below:

shop	short term	long term
1 electrical		✔
2		

After mapping functions in Bayeux the pupils compared them with those in Ver-sur-Mer. There are more specialist shops in Bayeux. People travel into Bayeux from the surrounding villages to use them. Pupils worked out the percentages of different types of functions. **C** is a double bar graph comparing Ver-sur-Mer with Bayeux.

Another interesting feature of shopping in Bayeux is the market (**D**). The weekly market is common in French towns. Stall-holders travel to different markets during the week. Local people bring in fresh produce from their farms.

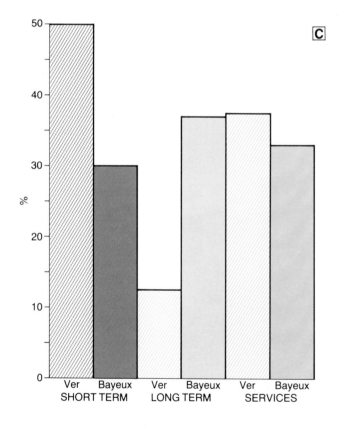

C

D

5 a How do the functions in Ver and Bayeux differ?

b Suggest reasons for the differences.

1.8 Catchment Areas

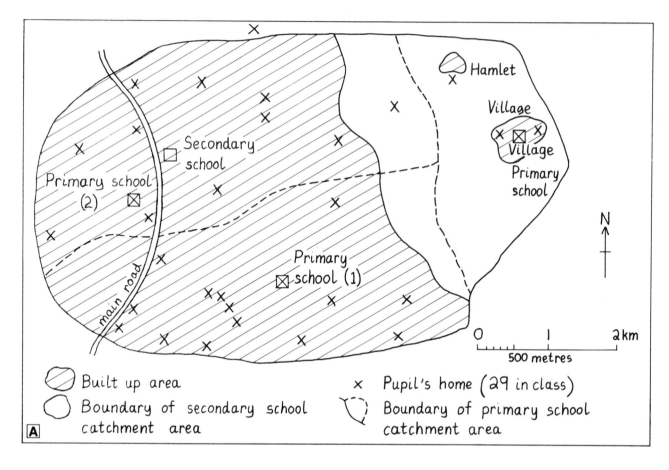

Built up area

Boundary of secondary school catchment area

× **Pupil's home (29 in class)**

Boundary of primary school catchment area

0   1   2km

500 metres

A

A shows the areas served by four schools. The 29 pupils in one class in the secondary school have marked their homes on the map. Two years ago the children went to three primary schools. All these are inside the **catchment area** of the secondary school.

Shops and services also have catchment areas. **B** shows the Ver-sur-Mer post office.

B

---

1 How many of the pupils live in the built-up area?

2 How many live within 1 kilometre of the secondary school?

3 How many of the pupils live outside the official catchment area of the secondary school?

4 How many children in the class went to primary school **1**, school **2** and the village school?

5 How far in kilometres is primary school **1** from the secondary school?

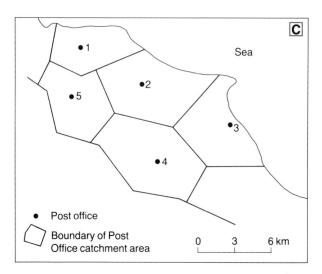

People living in the village and the surrounding farms use this post office – they are in its catchment area. **C** shows the catchment area of post offices in a rural area like the Normandy coast.

The pupils who visited France (see Units 1.6 and 1.7) were impressed by the huge hypermarkets. These very large shops are like giant supermarkets. They have large catchment areas which cover whole towns and several villages. The pupils saw hypermarkets at Caen (**D**) and Cherbourg (**E**).

6 Why do you think people are prepared to travel a long way to shop at a hypermarket? Is it because of the prices, the variety of goods, the car parking, or some other reason?

7 How do you think the catchment area of a hypermarket could be measured? What questions could you ask customers at a hypermarket in order to measure its catchment area?

FURTHER WORK

● In groups measure the catchment area of a local shop – or hypermarket, if you live near one. You can use the survey sheet in the *Activity pack*.

Rules for doing a survey
1 Do not do the survey in a big group
2 Tell the shopkeeper what you are doing
3 Be polite
4 Say where you are from, explain what you are doing, and state that you do not want the customer's full address – only the road or area
5 Ask as many shoppers as possible, then you will get a better result

● Ask people where they live, as they leave the shop. Plot the addresses on a large-scale map (such as a 1:10 000 or 1:2500) and draw in a catchment area. Present your work neatly and explain what you have done. Use the following headings:
My aim (what I wanted to do)
How I collected the data
How I presented the data
What I discovered or learnt
My conclusion
Problems I had doing the work

1.9 Paris – The Central Place

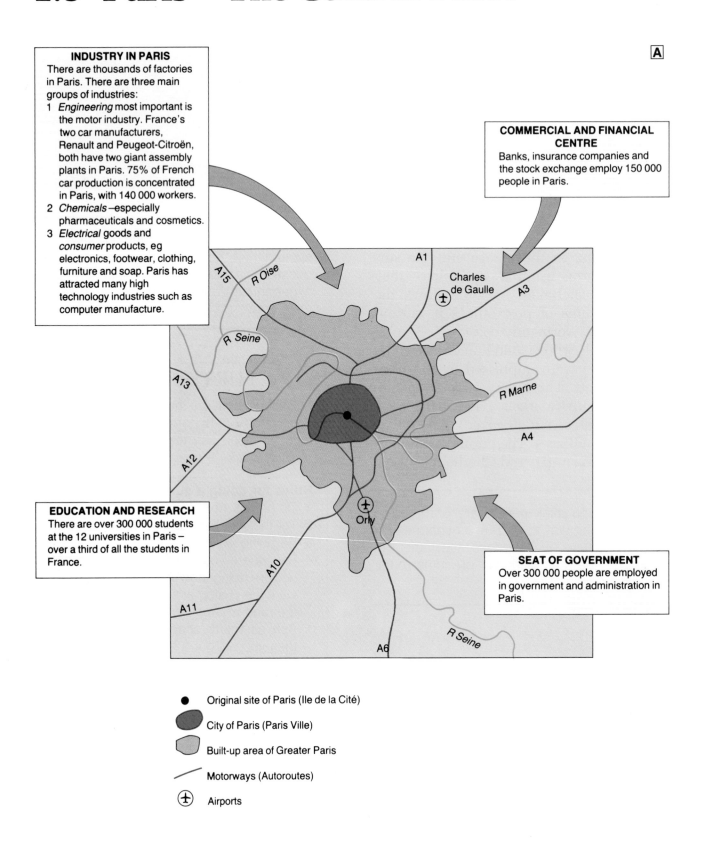

A

INDUSTRY IN PARIS
There are thousands of factories in Paris. There are three main groups of industries:
1 *Engineering* most important is the motor industry. France's two car manufacturers, Renault and Peugeot-Citroën, both have two giant assembly plants in Paris. 75% of French car production is concentrated in Paris, with 140 000 workers.
2 *Chemicals* –especially pharmaceuticals and cosmetics.
3 *Electrical* goods and *consumer* products, eg electronics, footwear, clothing, furniture and soap. Paris has attracted many high technology industries such as computer manufacture.

COMMERCIAL AND FINANCIAL CENTRE
Banks, insurance companies and the stock exchange employ 150 000 people in Paris.

EDUCATION AND RESEARCH
There are over 300 000 students at the 12 universities in Paris – over a third of all the students in France.

SEAT OF GOVERNMENT
Over 300 000 people are employed in government and administration in Paris.

A1
A15
R Oise
Charles de Gaulle
A3
R Seine
A13
R Marne
A12
A4
Orly
A10
A11
A6
R Seine

● Original site of Paris (Ile de la Cité)

City of Paris (Paris Ville)

Built-up area of Greater Paris

Motorways (Autoroutes)

✈ Airports

A **central place** is a settlement which provides services for its surrounding area. Villages, towns and cities are all central places. The largest central place in Europe is Paris. 9.3 million people live in the city. Paris is the capital of France.

Paris grew up in prehistoric times on an island in the river. This was an excellent defensive site which became an important river crossing. **A** shows how Paris has grown to become the most important city in France. Today, Paris is:

- almost eight times larger than the next largest French city (see **B**)
- the centre of the government of France
- the headquarters of all but a handful of the major French companies
- the most important centre of manufacturing industry in France
- the centre of French finance and commerce
- the focus of French transport and communications (**C**)
- the educational centre of France
- the cultural, artistic and entertainment centre of France

---

1 a Name the island on which Paris originally grew up (see **A**).

b Name the river in which the island is located.

2 Using your atlas to help you, study **A** and say which motorways lead:

a from Paris to L------.

b from Paris to M-----.

c from Paris to Ly-----.

d from Paris to B-----.

e from Paris to R-----.

f from Paris to C-----.

3 a What is the population of Paris?

b Name the three main groups of industries in Paris.

---

c How many students are there in Paris?

d How many people are employed in banks, insurance, the stock exchange, government and administration?

4 Table **B** shows the population of the eight largest French cities:

a Devise another way of showing this information (a bar graph; a pie graph; some other form of presentation).

b How many French cities have a population of over half a million?

City	Population (millions)
Paris	9.3
Lyons	1.2
Marseilles	1.1
Lille	0.9
Bordeaux	0.6
Toulouse	0.5
Nantes	0.5
Nice	0.5

B The population of France's major cities

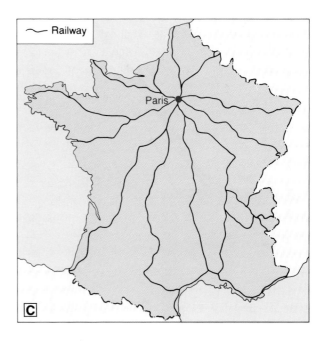

— Railway

C

1.10 Inner Cities – Land Use

You are visiting Paris for the weekend and you want to see all the famous sights. Your map shows the inner city area, with some of the most important buildings and monuments (**A**).

A

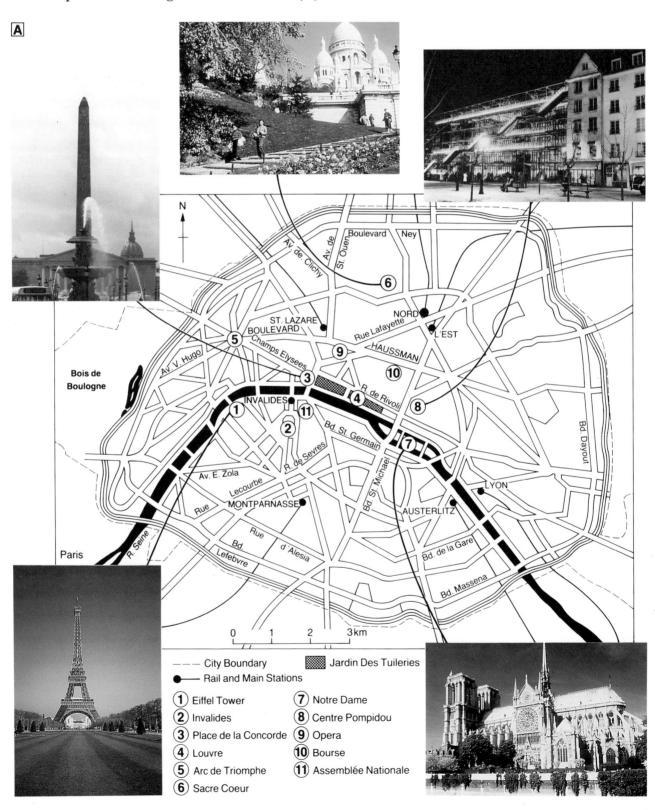

```
          0    1    2    3km

- - -  City Boundary          ▦  Jardin Des Tuileries
●━━  Rail and Main Stations

① Eiffel Tower              ⑦ Notre Dame
② Invalides                 ⑧ Centre Pompidou
③ Place de la Concorde      ⑨ Opera
④ Louvre                    ⑩ Bourse
⑤ Arc de Triomphe           ⑪ Assemblée Nationale
⑥ Sacre Coeur
```

Skills	using street map, measuring distance, designing pamphlet
Concepts	tourist attractions, overcrowding, slum clearance
Issues	inner city problems

1 You arrive at the Gare de Nord. You decide to walk to the Place de la Concorde.

a Along which road will you start walking?

b You cross the Boulevard Hausmann. Which famous building do you walk past?

c You arrive at the Place de la Concorde and walk through the Jardin des Tuileries. Which famous building is in the Jardin?

d You return to the Place de la Concorde and decide to walk to the Arc de Triomphe. Along which road do you walk?

e How far have you walked since leaving the Gare de Nord: 2 km? 5 km? 8 km? 12 km?

Inner Paris contains many buildings of national importance, including:

● The French Parliament (Assemblée Nationale)

● The palace of the President of France (Elysée Palace)

● Most of the offices and ministries of the government of France

● The supreme court (Palais de Justice)

● The national library (Bibliothèque Nationale)

● The Bank of France (Banque de France)

● The French stock exchange (the Bourse)

● The most important museums and art galleries in France including the Louvre

● The Pompidou Centre, a new art and cultural centre

Most of these buildings are on the tourist trail. Paris is a major **tourist centre**. It is easy to fall under Paris' spell when sitting at a pavement cafe and watching the hustle and bustle. Many think Paris is the most beautiful capital city in the world. But there is another side to Paris which most tourists do not see. The inner suburbs of Paris are largely old and overcrowded. Many homes lack basic facilities such as bathrooms and toilets – over a quarter of Paris homes have no indoor toilets. The 'East End' of Paris contains large slum areas and old factories and workshops.

There is little open space and few services. This is the Paris of the poor, the working class and the unemployed. Central Paris has over twice the population density of central London.

Inner Paris is changing rapidly (see **B**).

SLUM CLEARANCE [B]

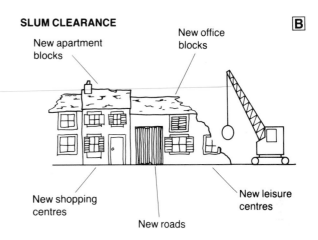

New apartment blocks

New office blocks

New shopping centres

New roads

New leisure centres

SLUM IMPROVEMENT

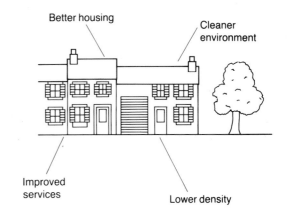

Better housing

Cleaner environment

Improved services

Lower density

2 Study **A**

a Name the famous church on the island in the River Seine.

b On which bank of the Seine is the Eiffel Tower (the river is flowing westwards through Paris)?

3 What are the inner suburbs of Paris like?

4 How is inner Paris changing?

5 Design a pamphlet to advertise the attractions of Paris to tourists. You should include drawings and a map of the major tourist sites.

1.11 Rich and Poor in Europe

Photographs **A–D** show different views in Europe. They were all taken during the last 10 years.

---

1 a Describe what you can see in each of the photographs **A–D** (mention the people, the clothes they are wearing, what they are doing, their surroundings).

b For each photograph, write a sentence about what you think the people's lives are like.

c Link the following statements with the four photographs:

This is hard work. I can't earn enough money to afford anything better.

Sure, conditions here are bad; but they are better than I could get at home.

I work hard and this is part of my reward. Why not?

I don't know what we'll do if the mine closes. Maybe we'll move away. Maybe we'll hope that something else turns up.

d Which of the people in the photographs would you most like to be? Why?

2 There are great contrasts between rich and poor people in Europe. The following things have some effect on whether a person will be rich or poor:

education   health   family background
race   type of job   where they are born
luck

For each factor in the list, write a sentence to say how it can affect a person's wealth.

Europe is a wealthy continent. Some countries are richer than others. One way of measuring the wealth of a country is to study the **GDP** (**Gross Domestic Product**). The GDP is the total money earned by a country through the production of goods and services. **E** shows the GDP per person of European countries.

3 Using your atlas and **E**, name four countries with a GDP per person of:
  a over $9000
  b between $6000 and $9000
  c between $3000 and $5999
  d under $3000.

4 a Name the only country in Western Europe with a GDP per person of under $3000.
  b In which part of Europe (north, south, east or west) are most of the poorest countries?

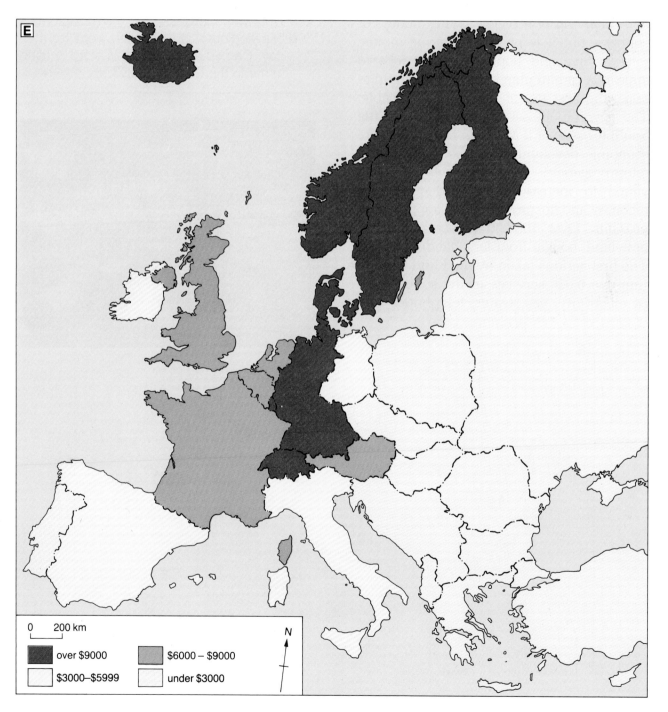

0 200 km	N

■ over $9000	▨ $6000 – $9000	□ $3000–$5999	□ under $3000

1.12 Government Planning Policies

In 1950 Southern Italy was one of the poorest regions in Europe. Nearly a million people were homeless. There was high unemployment. People were leaving, or **emigrating** from, Southern Italy at the rate of half a million a year. Conditions were so bad that the Italian Government stepped in to help.

How can a government improve the standard of living of a region? With money, of course – but it is no good pouring money into an area and hoping for new jobs to appear. A careful **plan** has to be drawn up to make the best use of the money available. The government set up the **Fund for the South** (*Cassa per il Mezzogiorno*), to run the economic development of the region. Land reform agencies were also set up. They aimed to increase farm production and incomes and to provide land for the poor peasants. Over 100 000 peasants received new land, bought from large landowners.

The first actions taken by the Fund were to improve basic services in the south, such as electricity, gas, water, telephones and roads. One fifth of all the Fund's spending between 1950 and 1970 was on such services. Without this, there was little hope of attracting modern industry into the region.

---

1 What evidence is given in the text that Southern Italy was a poor region in 1950?
2 a What was the Cassa per il Mezzogiorno?
  b 'Mezzogiorno' is the Italian word for mid-day. Can you think why this word is used to mean the South of Italy?

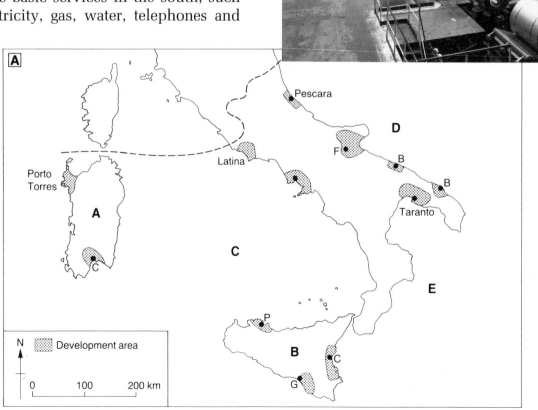

The government gave money from the Fund to companies who would build new factories in the South. It decided to concentrate the money in a number of favoured places called **Development Areas (A)**. Each Development Area was intended to grow rapidly and attract more jobs into the surrounding area, so that the whole region was helped. The Government gave the process a start by forcing state-owned industries to build new factories in the South. Oil refining, chemicals, engineering, iron and steel works were set up there.

---

3 Copy **A** and use your atlas to help you name:
  a the towns shown by their first letters
  b islands A and B
  c sea areas C, D and E
4 a Name six Development Areas in Southern Italy.
  b What was the aim of a Development Area?
  c Why have many oil refineries, chemical plants and steelworks been built in the South?

One of the most successful Development Areas is Taranto. Its deepwater port gives Taranto a great advantage. Europe's largest iron and steelworks has been built there (**B**) with a workforce of 17 000. Giant bulk carriers bring in coke from the USA and iron ore from Brazil and Africa. Several other industries have been attracted to Taranto, including metal engineering, oil refining, cement and washing machines. The new jobs have caused Taranto to grow quickly and a new town has had to be built to house the workers.

Unfortunately Taranto's success has not been repeated in many places. Southern Italy remains poor compared with the rich north. Unemployment is twice as high, average income half as much and emigration three times as great in the South. The government ended the Fund in 1984. 34 years of effort and £18 000 million have achieved a lot, but plenty remains to be done. Current government policy is to concentrate on agriculture, tourism and industry, such as food processing, based on local resources. In 1986 the government set up **The Agency for the Promotion of the Development of the South**. This is almost the Fund under a new name. It is a clear indication that the South's problems are far from over!

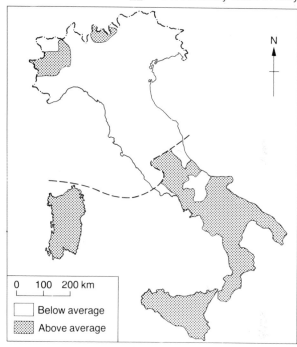

C Infant mortality rates in Italy

0   100   200 km

☐ Below average
▨ Above average

---

5 The table below shows the population of Taranto between 1951 and 1981:

Year:	1951	1961	1971	1981	1984
Population (thousands)	167	194	225	247	243

  a Draw a line graph for these statistics.
  b During which period did Taranto grow fastest?
  c What do you think was the main cause of the growth of population?
  d What has happened to the population since 1981? Why do you think this is?
6 a What does **C** show?
  b Why does **C** suggest that more needs to be done to help the south of Italy?
  c 'Current policy concentrates on agriculture, tourism and industry based on local resources.' Why do you think this is?

1.13 The rich world beyond Europe

Where do you think photographs **A**, **B**, **C** and **D** were taken? Photographs like this could be taken in many countries of the rich world. Map **E** shows which countries are part of the rich, or **developed**, world. In these countries, industry and farming are well developed. People have, on average, a high **standard of living**. In general, people are well educated. They eat well, have good medical services, and live a long time. Far fewer children die within their first year in developed countries than in the poor or **developing** world. Population in the developed world is only growing slowly.

The **gap** between the rich world and the poor world is getting wider. It is not only a gap of income but a gap of technology and power (**F**). It is also a gap of education, transport and services.

Skills	describing/sketching from photograph, line sketches
Concepts	rich world, rich/poor gap
Issues	world inequality

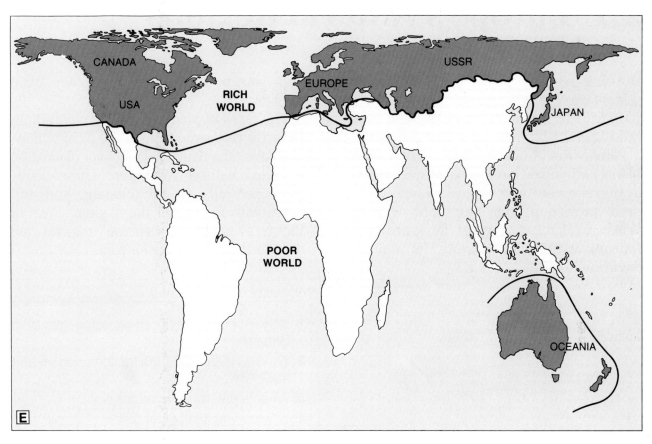

CANADA

USSR

USA

EUROPE

JAPAN

RICH WORLD

POOR WORLD

OCEANIA

E

1 **a** Briefly describe what each of photographs **A–D** shows.

   b Why could similar photographs be taken anywhere in the rich world?

2 **B** shows the port of Long Beach, Los Angeles. Write a short paragraph about the port. Include the following words and phrases in your description: reclaimed from the sea; large sheltered docks; container areas; warehouses.

3 **a** Sketch some of the housing in photograph **C**.

   b Label the main features on your sketch, eg high rise apartments, small balconies.

   c Describe the type of housing shown by your sketch.

4 Study **E**. What countries outside Europe are part of the rich world? Use an atlas to help you name these countries.

5 Look at **F**. Draw your own diagram to show features of the rich world. Use similar **line sketches** to illustrate your statements. You can only use two of the facts in **F**. (You do not need to include statistics.)

The Rich World F

25% of the world's people live in the **RICH WORLD**

They have. . .

83% of the world's income

70% of the world's food grains

89% of the world's education spending

95% of the world's science and technology

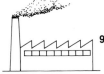
92% of the world's industry

94% of the world's health spending
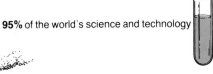

85% of the world's spending on military equipment

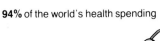

2.1 Europe's Natural Regions

Did you see any plants on your way to school today – wild plants, plants in people's gardens..? Perhaps some of them were indoors. Which plants grow best indoors?

Plants from tropical places, like rubber plants and palms, have to be grown indoors as they are used to a tropical climate. They need protection from our cold winters. When we protect a plant by growing it indoors we try to recreate its **natural environment**.

A shows the general needs for plant growth.

The types of plants that grow naturally in the wild vary, according to the local climate and soils. The natural plant cover of a region is called **natural vegetation**. These plants grow naturally in the climatic and soil conditions found in the region. Five of Europe's natural vegetation regions are described on the opposite page.

A

Atmosphere

Light energy

Water

Carbon dioxide

Soil

Water & Nutrients

The atmosphere gives light energy, carbon dioxide and water. The soil provides nutrients as dissolved mineral salts. In the leaves, water mixes with carbon dioxide, using sunlight to produce food and give off oxygen. This process is called **photosynthesis**. Surplus water is passed into the air from the plant's leaves. This process is called **transpiration**. Plants will adapt to changes in temperature and water supply, and so vary in shape and size. Generally they will not grow at temperatures below 6°C.

1 Why is it difficult to grow some plants in Britain?
2 What do plants get from the atmosphere and soil?
3 What is natural vegetation?
4 Why do you think most plants in Britain stop growing in winter?
5 What natural vegetation grows around or near your school?
6 What type of vegetation is found in the far north of Europe?
7 Give two differences between deciduous and coniferous trees.
8 How does the climate affect plant growth in the Mediterranean scrub region?

FURTHER WORK

● The natural vegetation of Europe has been changed by people. Over hundreds of years they cleared forests for timber and fuel, ploughed grasslands to grow food and drained marshes to make useful land.

Use the *Index* and look through this book to find other ways in which people have affected the natural vegetation of Europe.

Skills	research, interpreting photograph
Concepts	vegetation/climate, natural vegetation, natural regions
Issues	natural vegetation as a resource

Mountain vegetation

Average temperature decreases with height. Generally, no plants grow above 4000 metres above sea level. The vegetation of mountain regions changes with height – deciduous trees grow at the bottom and above them coniferous trees. These then give way to small plants, shrubs and finally alpine plants. Soils become thinner as slopes get steeper.

Coniferous forest

Coniferous trees have straight trunks and needle-shaped leaves. They are shaped to withstand high winds and the snowfall of long, cold winters. The trees are evergreen (they have leaves throughout the year). Spruce and pine are examples of trees found in coniferous forest areas.

Mediterranean scrub

In this region summer heat is overpowering and loss of water is high. Low trees with waxy leaves grow here. The trunks have thick bark and appear rough and twisted (eg cedar). The trees store winter rainfall in trunks and leaves, for the summer. In drier parts, sweet-smelling herbs and shrubs grow (eg lavender, rosemary and thyme).

Tundra

Snow covers these cold lands in winter. Below the surface soil the ground is permanently frozen. The topsoil thaws during the short growing season, when grasses, quick-flowering plants and mosses can grow. Trees are rare, and only dwarf varieties can survive.

Deciduous forest

The trees of the deciduous forest shed their leaves during winter when temperatures generally fall below 6°C. They need a growing season of at least five months. Examples of deciduous trees are oak, beech, birch and ash.

35

2.2 Europe's Climate

In Unit 2.1 you saw how natural vegetation is affected by climate. What is climate? **Climate** is the annual pattern of weather (ie rainfall and temperature changes) for an area. Weather is recorded over 30 years to find what it is like in an average year. When someone wants to describe a climate they usually give the average monthly temperatures and rainfall.

Geographers are very interested in studying the climate of an area because this can help them understand environments and the activity of people.

Discuss with your group how Britain's climate affects you:

How do you feel when it rains? ...when it snows? ...when it is sunny and hot?

Do you prefer winter or summer?

---

1 What is climate?

2 Why are geographers interested in climate?

3 What is the climate called within the Arctic Circle?

Geographers often use graphs to represent climate figures for a place. It is easy to see changes in temperature and rainfall throughout a year on a graph. The wettest/driest or hottest/coldest parts of the year can be quickly discovered.

Climate graphs

Europe has a number of different climates (see **B**). **C** shows **climate graphs** for four towns, each with a different type of climate.

Geographers use certain **conventions** for climate graphs: the 12 months of the year are shown along the bottom axis. A line graph is used for temperature and a bar graph for **precipitation** (rainfall, snow, hail, sleet).

a As you go north in Europe the effect of the sun is reduced and so the temperature drops.

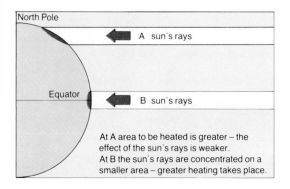

At A area to be heated is greater – the effect of the sun's rays is weaker.
At B the sun's rays are concentrated on a smaller area – greater heating takes place.

b If you walk up a mountain the temperature will drop by 1°C for every 150 metres of height gained. The air also becomes thinner; climbers need oxygen on high mountains.

c The sea takes longer to warm up than the land, and cools down more slowly. This is why it is always cooler in the sea than on the beach on a summer's day. The temperature of the sea affects coastal areas; in winter they are slightly warmer than inland regions; in summer they are slightly cooler.

d Places far from the sea tend to be hot in summer and cold in winter.

e Ocean currents can help keep coasts warm in winter if they flow from warmer parts of the ocean, or colder if they flow from cooler ocean areas. For example the warm North Atlantic Drift (see **B**) keeps western Britain warmer in winter.

f The **prevailing** wind is the most usual wind. We generally describe winds by the direction from which they blow (see **B**). The temperatures and moisture brought by the wind depend upon the climate of the region it blows from. For example westerly winds from warmer lands cross over the Atlantic and bring rain to western Britain.

A

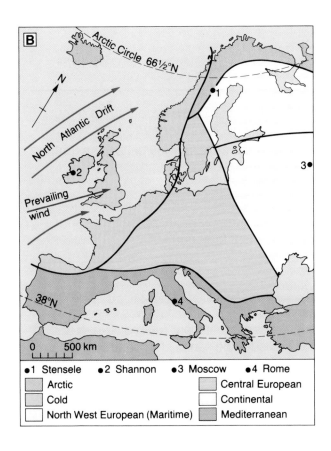

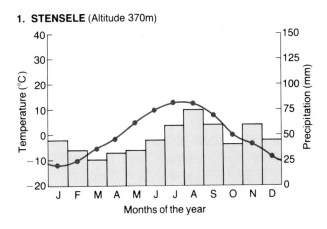

1. STENSELE (Altitude 370m)

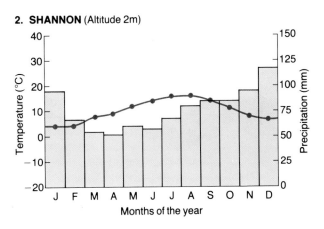

2. SHANNON (Altitude 2m)

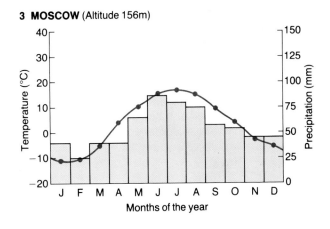

3 MOSCOW (Altitude 156m)

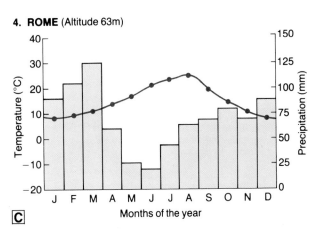

4. ROME (Altitude 63m)

B

- ●1 Stensele ● 2 Shannon ● 3 Moscow ● 4 Rome
- Arctic
- Cold
- North West European (Maritime)
- Central European
- Continental
- Mediterranean

4 What type of climate is found in Southern Europe?

5 Will it rain (or snow) in Moscow in January, February, March, November and December? Give reasons for your answer.

6 **Maritime** means influenced by the sea. The North Atlantic Drift (see **B**) is a warm ocean current that flows past north west European coasts. How do you think this modifies the climates of this region?

7 **Annual range of temperature** is the difference between the highest and the lowest monthly temperature in a year.
   a Which of the four graphs in **C** shows the greatest annual range of temperature?
   b Why do you think this place has such a wide temperature range (use **A** to help you)?

8 'Warm, wet winters, hot, dry summers'. Which of the climates in **C** does this describe?

9 Imagine that you are writing to someone about the climate of Stensele. Use **C** to describe the climate.

C

37

2.3 Rocks

It is easy to forget that **rocks** are close beneath our feet. Usually they are covered by soil (which partly consists of broken rock), plants, buildings or roads. **Geology** is the study of the structure and history of the Earth. It includes the study of rock.

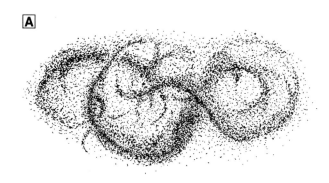

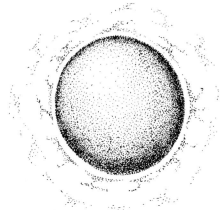

1 How old is the Earth, according to popular ideas of how it was formed?

2 What is the popular idea of how the Earth was formed?

3 Study **B**
  a How thick is the crust under oceans?
  b What is the diameter of the Earth from continent to ocean?

The Earth is part of the Solar System. No-one is certain how the Solar System was formed. One popular idea is that many millions of years ago, the sun formed at the centre of a hot, whirling cloud of dust and gas. Other parts of the cloud split off and became planets.

At first the Earth was still very hot. Gradually the surface cooled down. Over 4600 million years ago a thin scum collected on the surface. Very slowly it hardened into a solid and formed the Earth's **crust**. This solid material is a type of **igneous** rock. The Earth's crust is very thin. At its thickest point it is only 50 kilometres thick. Inside the crust, the Earth is very hot.

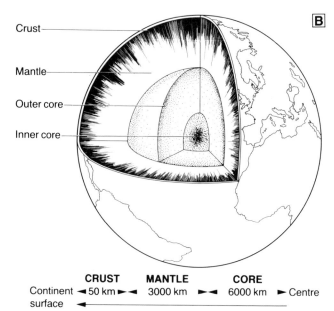

Crust

Mantle

Outer core

Inner core

CRUST	MANTLE	CORE	
Continent ◄ 50 km ►◄	3000 km	►◄ 6000 km	► Centre
surface ◄			

The **crust** is only 5 kilometres thick under the oceans and up to 50 kilometres thick under the continents.

The **mantle** is so hot that much of the rock has turned to liquid. It contains hot, liquid rock called **magma,** at a temperature of over 2000°C

The outer part of the core is liquid, the inner part is solid. Its temperature is 2500°C.

In the beginning

There are three basic groups of rock. **Igneous** rock (**C**), **sedimentary** rock (**D**) and **metamorphic** rock (**E**).

As the surface of the Earth's crust was eroded and creatures began to live and die in the sea, Sedimentary rocks were formed. These are formed from tiny particles (for example, of rock) that have been carried and then deposited by water, wind or moving ice. Over millions of years the deposits turned to rock.

Igneous and sedimentary rocks can be **compressed** by movements in the Earth's crust (for example folding) or **baked** by magma making its way to the surface. For example shale (sedimentary rock) can be **changed** into slate (metamorphic rock)

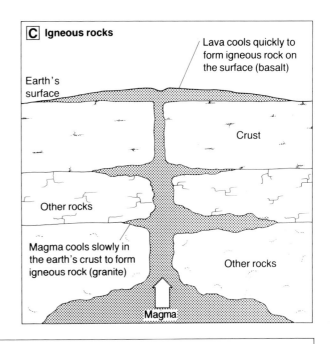

C Igneous rocks

Lava cools quickly to form igneous rock on the surface (basalt)

Earth's surface

Crust

Other rocks

Magma cools slowly in the earth's crust to form igneous rock (granite)

Other rocks

Magma

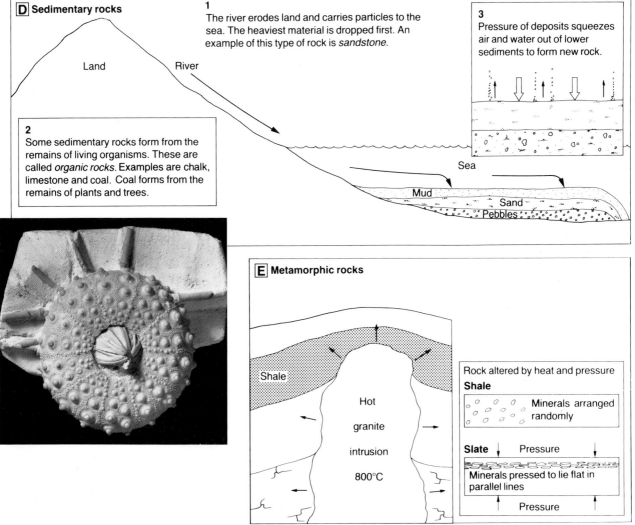

D Sedimentary rocks

Land

River

1
The river erodes land and carries particles to the sea. The heaviest material is dropped first. An example of this type of rock is *sandstone*.

3
Pressure of deposits squeezes air and water out of lower sediments to form new rock.

2
Some sedimentary rocks form from the remains of living organisms. These are called *organic rocks*. Examples are chalk, limestone and coal. Coal forms from the remains of plants and trees.

Sea

Mud

Sand

Pebbles

E Metamorphic rocks

Shale

Hot

granite

intrusion

800°C

Rock altered by heat and pressure
Shale

Minerals arranged randomly

Slate Pressure

Minerals pressed to lie flat in parallel lines

Pressure

39

2.4 What ice did to the land

In the last 2 000 000 years there have been several periods when ice covered large parts of the world's surface. During these periods summers were like our winters; in winter the temperature was below 0°C and the snow and ice never thawed. Between the periods of ice cover there were warmer periods when the ice retreated (**A**).

The last **Ice Age** ended about 10 000 years ago. At some stage during this period nearly one third of the world was covered with **ice sheets** over 100 metres thick. Valleys were filled with rivers of ice called **glaciers**.

Today an ice sheet covers Greenland and there are still glaciers in the Alps and Europe, in the Norwegian mountains, in the Rocky Mountains of North America and on Mount Cook in New Zealand.

B shows a valley glacier near Mount Cook. Valley glaciers like this gradually move downhill, due to the weight of ice. When the

temperature rises the glacier halts and begins to melt. If the temperature remains above freezing for any length of time the edges of the glacier melt away and it **retreats** back up the valley.

C shows how a glacier is formed.

The work of ice

As a glacier moves it erodes the bottom and sides of its valley. The ice freezes to the rock and tears (or **plucks**) some of it away. Weathered rock is carried along by the glacier and acts like a huge file, wearing away the surface underneath – a process known as **abrasion**. The material **trans-**

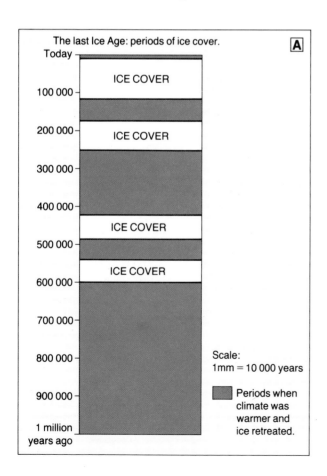

The last Ice Age: periods of ice cover.

Scale:
1mm = 10 000 years

Periods when climate was warmer and ice retreated.

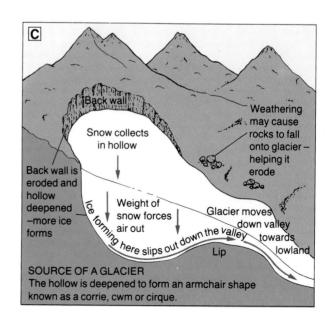

SOURCE OF A GLACIER
The hollow is deepened to form an armchair shape known as a corrie, cwm or cirque.

Skills	time line interpretation, research
Concepts	ice ages, glacier, erosion, deposition
Issues	use of glaciated landscape

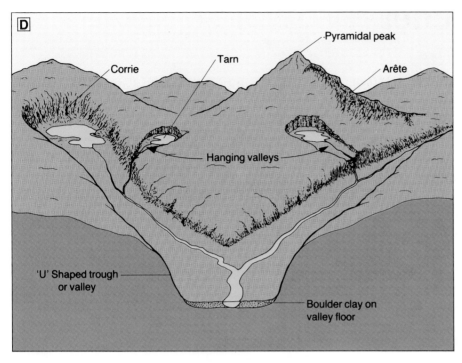

Features of highland glaciation

The **'U' shaped valley** has a flat floor with steep sides. The glacier has eroded an old river valley and left tributary valleys 'hanging' above the main valley. Water has filled part of the **corries** to form lakes called **tarns**. A knife-sharp ridge has been left which is called an **arête**. Where **three** corries back each other a **pyramidal peak** is formed. Rivers have started to work on the land, eroding the boulder clay and depositing new material on top of the glacial deposits.

ported by the glacier is called **moraine**. Some of this material is **deposited** on the sides and floor of the valley, the rest is deposited at the front, or **snout,** of the glacier when it halts. Ice sheets act in a similar way, but on a much larger scale.

Effects on the land

Although today only about a tenth of the earth is covered by ice, we can still see a lot of evidence of how ice changed the land by **erosion**, **transportation** and **deposition**. Both highland and lowland areas are affected by the work of ice. **D** shows some of the features produced by the work of highland glaciers.

Ice sheets can round hills, create hollows, smooth and polish outcrops of harder rock and deposit material as boulder clay.

Effects on people

● Boulder clay is very fertile; areas once covered by ice are ideal for agriculture (eg East Anglia).
● In highland areas (eg in Scandinavia) the flat valley floors carved out by glaciers are used for settlement, routeways and farming.
● Steep, glaciated valleys can be dammed to create reservoirs for water supply or for generating hydro-electric power (eg Norway and North Wales)
● The dramatic scenery created by ice attracts tourists (eg the Alps in Europe, the Southern Alps in New Zealand).

1 What is an Ice Age?
2 Look at **A**
  a How many periods of ice cover were there during the last Ice Age?
  b Use the line scale to measure how long each period lasted.
3 Where can you find glaciers today?
4 Study **C** and explain how a glacier is formed.
5 What makes a glacier retreat?

FURTHER WORK

● Imagine what would happen if there was another Ice Age. Draw a picture, or write a story, to show how it would affect life in your area.

2.5 Avalanche!

Each year an average of 25 people are killed in the Swiss Alps by **avalanches** – the 'white death' (**A**). An avalanche is a frightening example of a **natural hazard**. It is a mass of ice and snow which hurtles down a mountainside. Speeds of over 300 km per hour have been reported.

What causes an avalanche? As snow builds up on the mountainside its weight increases. It can become overloaded and unstable. The avalanche can be triggered by an earth tremor, by the wind, by rapid increases in temperature, or even by a single careless skier. As the avalanche careers downhill it forces a blast wave of air before it which can be powerful enough to shatter buildings. Avalanches kill and injure people and animals, demolish buildings and smash forests. They can block roads and railways. They can cause floods by blocking rivers.

The number of deaths caused by avalanches has increased rapidly in recent years (**B**). This is due to the increasing number of tourists who visit mountain areas on skiing holidays. Most of the more popular ski slopes are safe, but there is a growing interest in cross-country skiing. Sometimes people may ski into dangerous areas without realising it.

B

Protection against avalanches

The earliest form of protection against avalanches was the planting of **shelter belts** of trees to break up and slow down the rushing snow. Such shelter belts are still planted today. Other forms of defence are also used (**C**). The snow bridge and snow drift structures are designed to make the snow more stable. Sometimes it may be better to encourage small avalanches to happen. Small explosions are used to set off minor avalanches, so that the damage can be controlled. The small avalanche stops the build-up of snow which might lead to a major disaster.

Avalanches have even been used as a weapon of war. During the First World War the Italian and Austrian armies fighting in the Dolomite Alps used artillery to set off avalanches that killed tens of thousands of men.

A

C Avalanche protection structures

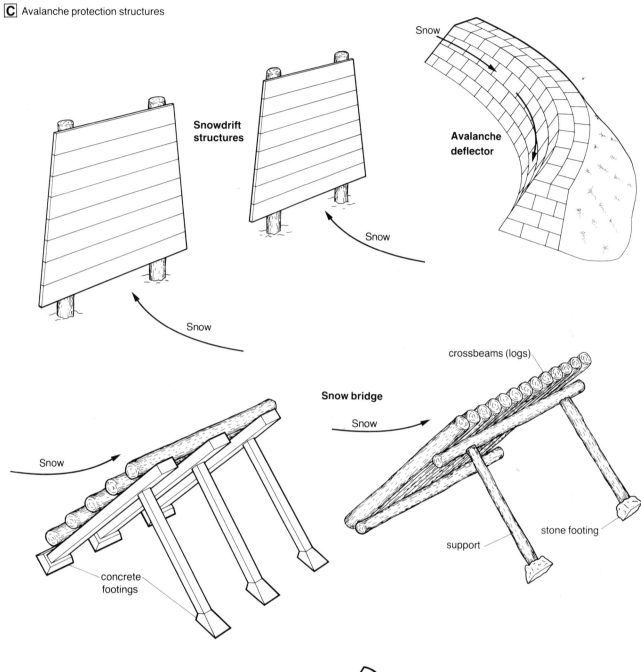

Snowdrift structures

Avalanche deflector

Snow

Snow

Snow bridge

Snow

crossbeams (logs)

stone footing

support

concrete footings

Snow

1 What is an avalanche?

2 What causes an avalanche?

3 Why has the number of deaths due to avalanches increased rapidly in recent years?

4 How can people protect themselves against avalanches?

5 You work for the tourist office of a Swiss **canton** (local authority). Design a poster warning people of the danger of avalanches.

FURTHER WORK

● You are skiing along the side of a mountain valley. On the opposite side of the valley you can see a village and a railway line nestling beneath a thin belt of trees. You hear a low whistle which increases rapidly to an awful howl. You stop skiing and look in horror at an avalanche hurtling down towards the village.

Write a report about what happened for your local newspaper.

2.6 Volcanic Activity

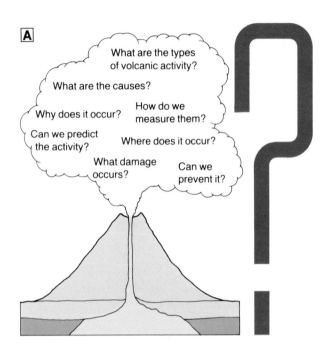

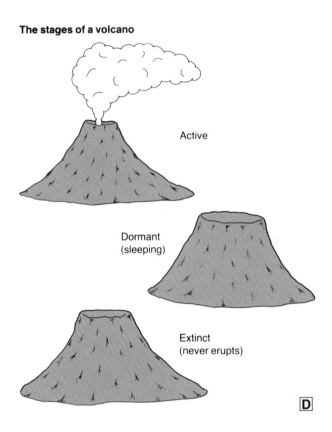

The stages of a volcano

Active

Dormant (sleeping)

Extinct (never erupts)

The types of volcanic activity

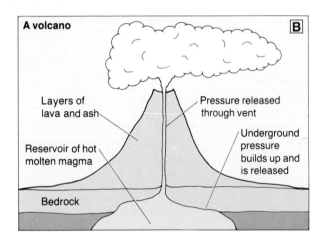

Volcanoes may look spectacular and exciting (**E**) but they can sometimes have tragic effects (**F**).

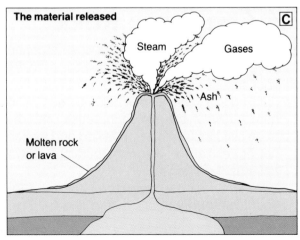

F

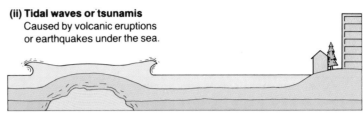

(ii) Tidal waves or tsunamis
Caused by volcanic eruptions
or earthquakes under the sea.

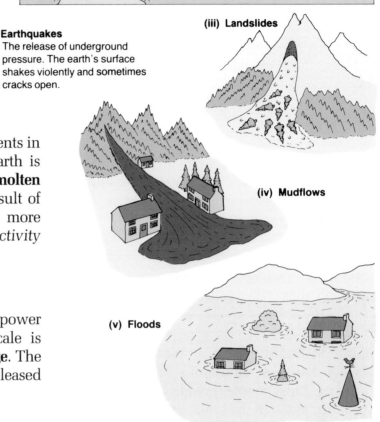

(i) Earthquakes
The release of underground
pressure. The earth's surface
shakes violently and sometimes
cracks open.

(iii) Landslides

(iv) Mudflows

(v) Floods

Volcanoes are associated with movements in
the Earth's **crust**. The crust of the Earth is
about 50 km thick (see Unit 2.3). The **molten
rock** beneath the crust moves as a result of
convection currents. (You can find more
about this in the *Resource and activity
pack*.)

Measuring earthquakes

There are two ways of measuring the power
of earthquakes (**G**). The **Mercalli** scale is
based on the amount of **visible damage**. The
Richter scale measures the **energy** released
in the earthquake.

MERCALLI intensity (degree of shaking)	**Description**	RICHTER magnitude (total energy released)
I	detected only by scientific instruments	2
II	feeble: felt only by sensitive people	3
III	slight: like vibration from a passing lorry	
IV	moderate: rocking of loose objects; parked cars shake	4
V	quite strong: felt by most people; church bells ring	5
VI	strong: most people frightened; windows break; items shaken off shelves	
VII	very strong: general alarm; walls crack	
VIII	destructive: car drivers find it difficult to steer; chimneys collapse	6
IX	ruinous: general panic; ground cracks, pipes shatter	
X	disastrous: large cracks open; many buildings destroyed; landslides	7
XI	very disastrous: most buildings, pipes, cables, bridges, dams etc destroyed; major land-slides and floods	
XII	catastrophic: total destruction; objects tossed into air; ground rises and falls in waves	8

G

2.7 Volcanic Disasters

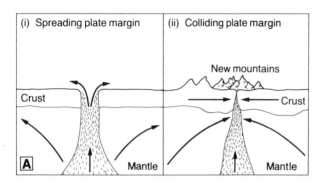

Volcanic activity occurs on the plate boundaries or **margins**. Here the plates meet and there are weaknesses in the Earth's crust. **A** shows two types of plate margin.

---

1 Study **A**. Describe how the plates are moving at the two types of margin.

Mount St Helens (**B**) is a volcano in the Cascade Range of mountains in Washington, USA. Most people thought the volcano was dormant – it had not erupted since 1857. Then, in March 1980, Mount St Helens began to wake up. **C** and **D** describe what happened.

C

20 *March* There was an earthquake, centred on the mountain.
27 *March* There was an eruption of ash and steam. A new crater formed, 50 metres deep.
28 *March* A second eruption sent ash over a distance of 3 km and started snow avalanches.
April Activity continued. A 100 metre bulge appeared in the side of the mountain.
18 *May* The volcano's biggest eruption took place. There was a big earthquake, and at 8.32am the bulge on the mountain began to collapse, followed by a landslide, a blast and a surge.

Stages of the Mt St Helens eruption

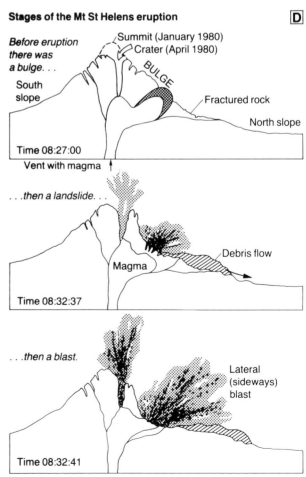

Skills	interpreting diagram, time line
Concepts	plate movements, volcanic hazards, prediction, prevention
Issues	managing disasters

This volcanic activity resulted from the movement of the Juan de Fuca Plate, which slid under the North American Plate (**E**).

Some effects of the eruption

● The main blast was of warm ash, quickly mixed with wood, rock and soil. All the fish and animals in the surrounding area were killed.

● Ash reached the Atlantic coast of the USA in three days. Within 17 days it had been carried all round the world in the atmosphere.

● There were log-jams over 30 km long on nearby rivers; floods and mudflows also occurred.

● Property damage was estimated at one billion dollars.

Prevention and prediction

Although 61 people died at Mount St Helens, the tragedy might have been much greater. Scientists were able to **predict** the volcanic activity. All the time the mountain was bulging it was being studied. Hazard warnings were issued whenever danger was near.

In other areas people have taken action to reduce volcanic damage. Barriers can be built on the sides of volcanoes to prevent landslides and mudflows. The volcanic eruption on the island of Heimaey, off southern Iceland, in 1973, could have cut off the fishing harbour. Firefighters hosed the advancing lava and cooled it. The harbour entrance was saved!

We cannot stop volcanic eruptions. All we can do is plan ahead and try to manage the effects of such natural hazards.

FURTHER WORK

● Draw a **time line** like the one below to show the different stages of the Mount St Helens eruption. Start with the March 20th earthquake. (Note: smaller eruptions continued throughout the whole of 1980.)

Time line

Earthquake	Ash
	Steam
	New Crater

March 20   March 27th

● Draw a diagram to show the different effects of the eruption.

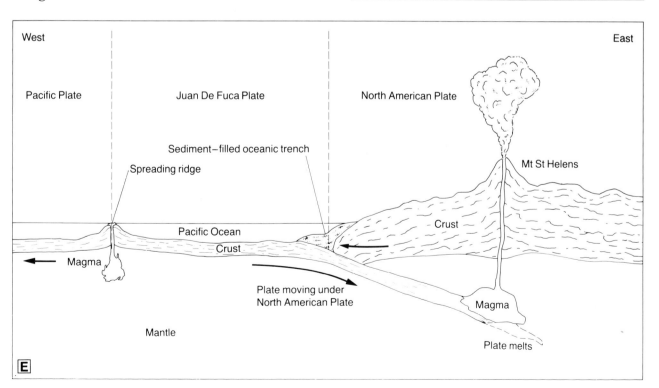

2.8 The Work of Rivers

Have you ever stood by a river or on a bridge and watched the water flow downstream? A river's size and speed depend on many factors, including climate, the rocks over which it flows, its stage of development (**upper**, **middle** or **lower course**) and how people use it.

What all rivers have in common is a **source** (usually highland) and a **mouth** (usually a lake, sea or ocean). On the way downstream from source to mouth many changes take place (**A**).

Geographers are interested in how rivers work. Rivers are often of use to people (for **transport**, **irrigation** and **water supply**) as well as sometimes being a natural hazard (**flooding**). Rivers drain the land and carry surface water to the sea. As they flow, they erode (**B**), **transport** and **deposit** (**C**) material.

The processes of erosion and deposition can be seen most easily at a bend or **meander** in the river (see **C**). The water erodes the outer curve and deposits material on the bed of the inner curve.

Rivers are at their most active after heavy rainfall. If you observe a river after a storm you may notice that the water becomes discoloured by the material it carries downstream.

---

1 Explain the meaning of the following words used in the study of rivers: source, mouth, bed, bank, meander, tributary, confluence, erosion, transport, deposition, load

---

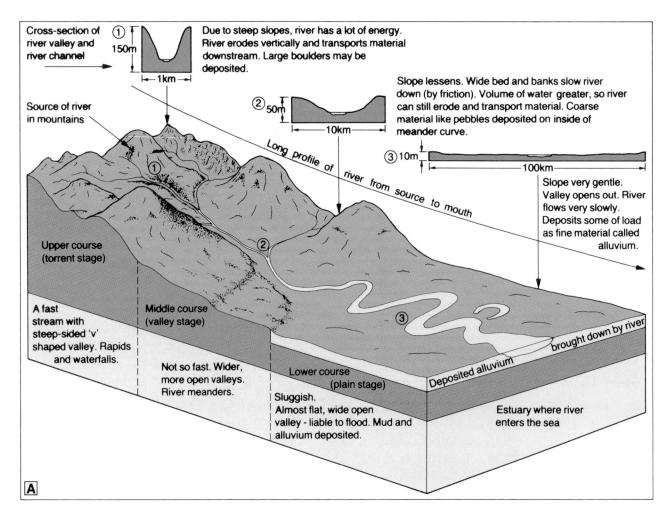

Cross-section of river valley and river channel

150m

1km

① Due to steep slopes, river has a lot of energy. River erodes vertically and transports material downstream. Large boulders may be deposited.

② 50m

10km

Slope lessens. Wide bed and banks slow river down (by friction). Volume of water greater, so river can still erode and transport material. Coarse material like pebbles deposited on inside of meander curve.

③ 10m

100km

Source of river in mountains

Long profile of river from source to mouth

Slope very gentle. Valley opens out. River flows very slowly. Deposits some of load as fine material called alluvium.

Upper course (torrent stage)

A fast stream with steep-sided 'v' shaped valley. Rapids and waterfalls.

Middle course (valley stage)

Not so fast. Wider, more open valleys. River meanders.

Lower course (plain stage)

Sluggish. Almost flat, wide open valley - liable to flood. Mud and alluvium deposited.

Deposited alluvium brought down by river

Estuary where river enters the sea

A

Skills	definition, interpreting diagram, research
Concepts	source/mouth, erosion/transport/deposition, long profile, cross-section
Issues	people's use of rivers

2 Copy and complete the following table, using information in **A**.

Stage of development	Speed of flow	Work of river	Features
upper middle lower	fast/torrent	erosion/transport	steep 'V' valley

Erosion occurs when a river wears away its channel. The erosion may be caused by a combination of: [B]

a the force of the water;

b the type of rocks over which the river flows (some rocks, eg limestone, dissolve in water);

c the action of small pieces of rock in the water, chipping away at the sides and bed;

d larger pieces of rock smashing together (forming the smooth, round pebbles often found on the river bed).

a Force of water

b Action of rock

c Large pieces of rock smash together

(i)   (ii)

3 Study **A**. Describe how the river channel (the banks and bed actually touched by water) changes shape during the three stages of development.

4 Why does rock material carried by running water become rounded to form pebbles?

The river **transports** or carries the material it has eroded downstream. The material being carried is called the **load**. [C]

When the river slows down it can no longer carry its load. It **deposits** the material on the bottom of its channel. It deposits the largest rocks and stones first (the **bedload**), then smaller and smaller material, until in its lower course it deposits very fine **alluvium**.

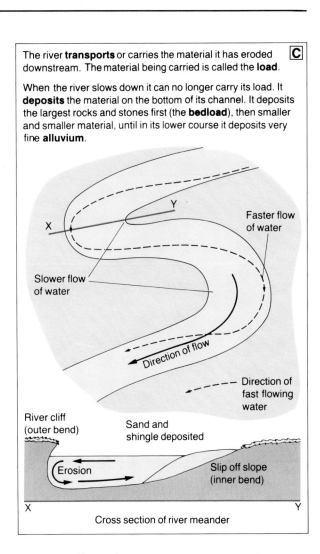

Cross section of river meander

5 Draw a diagram to illustrate how erosion and deposition take place at a meander.

6 How does the size of load being deposited change through the three stages ?

7 Use keywords from this unit to make a wordsearch on the work of rivers.

8 Look through this book and list:
a how people make use of rivers
b how people's actions affect rivers.

2.9 Controlling the Rhône

A shows a glacier high in the Swiss Alps. The melting water of this glacier forms one of Europe's longest rivers, the Rhône (see **B**). The river flows over 800 km from its **source** in Switzerland, through France, to the Mediterranean Sea. Before entering France the river flows through Lake Geneva. Then it winds along a series of deep valleys through the ranges linking the Alps and the Jura mountains. At Lyons the Rhône is joined by its largest tributary, the Saône, and turns southwards towards the still-distant sea.

1 **a** Where does the River Rhône rise?
  b Describe the course of the River Rhône between its source and the sea.
2 Why was the River Rhône 'a threat' to people in the past?
3 What do the following words mean?
  a source of a river **b** floods **c** drought **d** navigation?

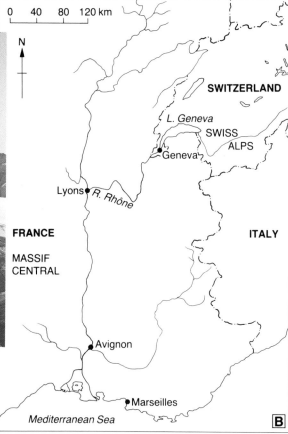

The valley of the River Rhône forms a narrow corridor between the mountains of the Massif Central and the Alps. This has long been an important routeway linking northern France with the south of the country. Yet until recently few people lived along the Lower Rhône valley. The river was dangerous. It flowed swiftly. In spring the melting snows of the Alps and Massif Central caused serious **floods** when the river's level rose and it overflowed its banks. In the summer the region's long dry spells, or **droughts**, caused the level of the river to fall too low for safe **navigation**.

Since the early 1950s the lower Rhône Valley has been the target of a massive development programme (**C**):

● Flooding has been stopped by dredging and deepening the river, building weirs and barrages.
● Nine new ports have been built along the river. Barge trains of up to 5000 tonnes can reach Lyons.

Skills	description from photograph and map
Concepts	source, floods, navigation, drought, hydro-electric power
Issues	natural hazards, people's use of rivers

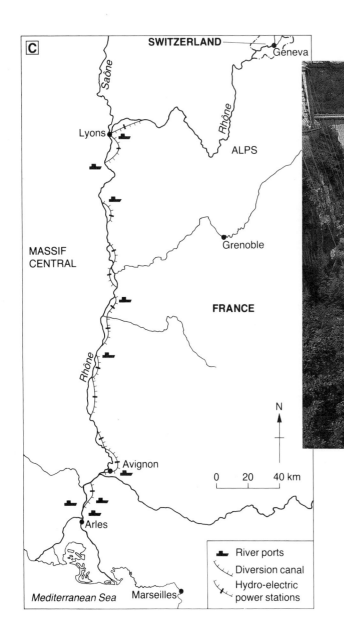

C

SWITZERLAND

Geneva

Saône

Rhône

Lyons

ALPS

Grenoble

MASSIF CENTRAL

FRANCE

Rhône

Avignon

N

0   20   40 km

Arles

Mediterranean Sea   Marseilles

River ports
Diversion canal
Hydro-electric power stations

D

- There are new industrial estates with chemical, cement and engineering works, oil refineries and food processing factories. These industries make use of the waterway for bulk transport and for cooling purposes.
- Twelve hydro-electric power stations have been built (**D**). These are sited on canals built to by-pass difficult stretches of the river. Barrages have been built, with locks for shipping to pass through. Turbines in the barrages are spun by the water. The turbines are linked to generators which produce electricity. Over 2400 megawatts of electricity are generated in total, that's enough for four million people!

- Water is supplied by irrigation canals to over 50 000 hectares of farmland. Fields which were once too dry to grow anything but wheat or olives can now grow fruit and vegetables. Farmers earn more money from apples, peaches and tomatoes than from wheat and olives.

The River Rhône, once a threat, is now controlled and safe. People are now moving into the Lower Rhône Valley, attracted by the new jobs in the area.

---

4 Why does the River Rhône no longer flood?

5 Why has the River Rhône attracted more industry?

6 How has farming been affected by the River Rhône development programme?

7 What advantages and disadvantages do you think barge transport has in comparison with lorry transport? (Think of fuel costs, pollution, speed, traffic congestion and cargo capacity.)

51

2.10 Flood Alert – Venice

Venice is one of the world's treasures. It seems to rise straight out of the waters of the Adriatic Sea. It is built on over 100 low islands in a lagoon sheltered from the open sea by a narrow sand bar called the Lido. Venice is a city of the sea. Canals are used instead of roads. Boats are used instead of cars. The buildings of the city date back many centuries to an age when Venice was the centre of a great empire spreading far and wide across the Mediterranean.

A trip through Venice is a memorable experience. The Grand Canal (**A**) is alive with boats. One beautiful palace follows another. There are countless churches, but the magnificent five-domed church of St Mark is the most famous. It dominates the elegant buildings of St Mark's Square (**B**). Museums, art galleries, shops, cafes and restaurants add to the city's attractions to the tourist.

But the unique beauty of Venice is under threat from pollution and from flooding.

Pollution

The petro-chemical works at Porto Marghera (see **C**) pollute the air and the sea. The **acid rain** (see Unit 2.12) causes the limestone and marble buildings of Venice to crumble.

Venice itself has a primitive sewerage system. Much of the city's sewage is poured into the lagoon, polluting the water.

Flooding

Flooding is the most serious problem facing Venice (**D**). Venice is at the head of the

Skills	description from photograph
Concepts	cultural heritage, tourism, acid rain, flooding
Issues	saving historical treasures

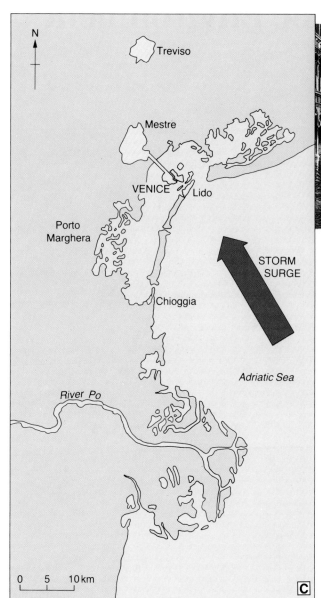

by pipeline from Alpine rivers. It is interesting to note that the rate of sinking had reached over 5 mm a year by the 1970s, but since the pumping was stopped the rate has fallen to just 1 mm a year.

Another factor in the flooding of Venice is the result of deepening the deepwater channel to Porto Marghera so that large, ocean-going ships can pass through the lagoon. Up to 7000 ships a year sail past and through Venice; the wash they create damages the ancient buildings.

Many of Venice's buildings suffer rising damp. Several families have abandoned the lower floor of their homes. Other buildings have been completely abandoned. The population of the old city of Venice has fallen from 175 000 in 1955 to only 65 000 by 1987.

Who will save Venice? There is much interest around the world in plans to save the city. Storm flood barriers like those built by the Dutch in their Delta Project could be the answer, but this would be very expensive.

Adriatic Sea, and is exposed to storm surges like those which threaten the southern North Sea and caused London to build its flood barrier (see Book 1 in this series). South-easterly winds pile up the water in the northern Adriatic, often flooding much of Venice. This is a natural hazard but the actions of people have added to Venice's problems.

Venice is slowly sinking! It is built on millions of wooden piles which are settling into the mud. The situation was made worse by the pumping of water from underground wells for the industries of Porto Marghera and Mestre. Protests stopped this in the late 1970s and the industries now receive water

1 a In which country is Venice?
  b At the head of which sea is Venice sited?
  c What is Venice built on?
2 Use **A** and **B** to help you describe the appearance of Venice.
3 How is Venice threatened by pollution?
4 a What is a storm surge?
  b How do the industries and port of Porto Marghera and Mestre affect Venice?

53

2.11 People Upset the Environment – The River Rhine

On 1 November 1986 a fire broke out at a warehouse in Basle, Switzerland. Firemen took two days to bring the fire under control because the warehouse was full of chemicals. This was serious enough, but one result of the fire was disastrous – 30 tonnes of poisonous chemicals flowed from the burning warehouse into the River Rhine. The chemicals were carried 700 km downriver to the North Sea, passing through some of the most densely populated areas of Europe.

Among the chemicals was mercury, a poison deadly to river life. Over half a millon fish were killed between Basle and Karlsruhe in West Germany. Countless birds, insects and river plants also died. This was the worst pollution disaster to hit the Rhine. But the river is always polluted. It has been called 'Europe's open sewer'!

The source of the River Rhine is in the snowfields and glaciers of the Swiss Alps. The water is clear and pure as it flows swiftly down the mountainside at the start of its 1200 km journey to the North Sea. The Rhine and its tributaries (smaller rivers joining the Rhine) flow through many of Western Europe's major industrial areas (**A**).

The waters of the River Rhine are in great demand for:

- **Cooling** purposes – many factories and power stations use the Rhine's water for cooling their machinery. This heats up the water. Warmer water holds less oxygen than cooler water. Fish die if the water temperature rises above 31°C.

- **Transport** – the Rhine is one of the world's most important inland waterways. Ocean-going vessels can navigate as far as Cologne, barge trains of 6000 tonnes can reach Mannheim and barges of 2000 tonnes can reach Basle (**B**). Fuel spillages are common.

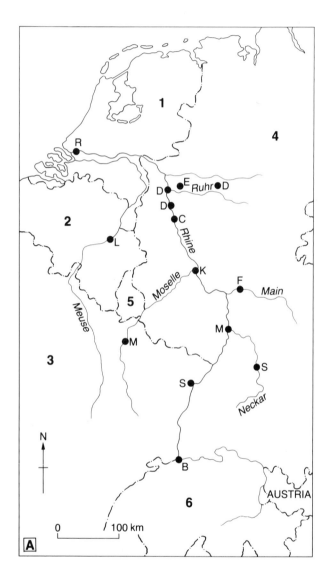

- **Drinking water** – the Rhine is a vital source of drinking water. The Dutch are most heavily dependent upon it. Over half of Dutch drinking water has to be taken from the river.

- **Waste disposal** – hundreds of factories, mines and towns pour their waste into the river. Chemicals and sewage are major pollutants.

Most of the chemicals that pollute the Rhine can be broken down and dispersed naturally by bacteria, but this takes time and reduces

Skills	atlas work, writing newspaper article
Concepts	river pollution, uses of water
Issues	pollution, international co-operation

the amount of oxygen dissolved in the water. Without oxygen, there can be no fish or plant life – the river dies.

Progress in cleaning up the Rhine has been slow. Six countries use the waters of the Rhine and its tributaries and they have not found it easy to agree who should pay for reducing the pollution. Perhaps the 1986 chemical disaster will speed up an agreement. Perhaps not.

1 a Where is the source of the River Rhine?
  b How long is the Rhine?
  c Through which countries does the Rhine flow?
  d What is a tributary?
  e Name four tributaries of the Rhine.
2 a Copy map **A** of the Rhine and its tributaries or use the map in the *Activity pack.*
  b Name the cities shown by their first letters.
  c Name the countries numbered **1** to **6**.
  d Use your atlas to help you mark on the points to which: ocean-going vessels; barge trains of 6000 tonnes; 2000 tonne barges can sail up the River Rhine.

3 Match up the following heads and tails:

Heads
Over half Dutch drinking water
Many factories use the Rhine's water
The Rhine has been called
Fish die if the water temperature

Tails
'Europe's open sewer'.
rises above 31°C.
for cooling their machinery.
has to be taken from the Rhine.

FURTHER WORK

● Write a newspaper article which describes the causes and effects of the 1986 pollution disaster on the Rhine. Think of a suitable eye-catching headline and use between 120 and 125 words.
● Do you think it is worth trying to clean up the Rhine? Who do you think should clean up the river?

2.12 Killer Rain

In some parts of Europe the forests are dying (**A**). This started to happen suddenly and without warning. In 1982 it was estimated that 8% of West Germany's trees were showing signs of damage from an unknown cause. The leaves turned yellow and dropped, growth was stunted, roots were shrivelled, the bark was damaged and in extreme cases the trees died. The mystery 'disease' advanced rapidly. By 1987 over half of West Germany's vast forests were dead or dying. This is not simply a German problem. 40% of Dutch forests are damaged. 35% of Swiss forests are dying.

What is causing this damage? The main suspect is **acid rain**. Acid rain is caused by air pollution (**B**). Sulphur dioxide and oxides of nitrogen are pumped into the atmosphere by power stations, oil refineries and motor vehicles. **C** shows the main sources of air pollution in West Germany.

These gases are chemically changed in the air. They turn into sulphuric acid and nitric acid. They mix with other chemicals to form acid rain.

---

1 What is acid rain?
2 Study **C**. Name the three main causes of
● sulphur dioxide; ● oxides of nitrogen
pollution in West Germany.

Acid rain does not simply destroy trees. It affects all living things. The rivers and lakes of Western Europe are becoming more and more acid. This acidity is killing fish and other animal and plant life. Especially badly affected is Scandinavia. By 1987 20 000 Swedish lakes were poisoned by acid rain, 4000 of them were so badly affected that nothing could live in them.

Some causes of air pollution in West Germany

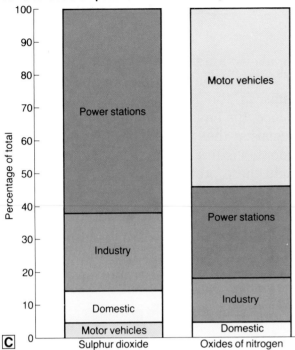

amount of sulphur dioxide pollution has fallen from 5.64 million tonnes in 1972 to 3.7 million tonnes in 1986, a 34% reduction. This is partly due to the decline of heavy industry in Britain. Oxides of nitrogen, however, have fallen by only 6% from 1.7 million tonnes in 1972 to 1.6 million tonnes in 1986.

Air pollution is an international problem. Britain remains Western Europe's worst producer of sulphur dioxide pollution. Only a third of this pollution falls on Britain, the other two-thirds falls elsewhere in Europe. 93% of the sulphur dioxide pollution in Norway comes from other countries, mainly Britain and West Germany. Much more needs to be done if the costs and damage of air pollution are to be reduced throughout Western Europe.

Acid rain also destroys stone and brickwork. Steel rusts five times more quickly in polluted areas. The damage is rapid. In Britain large buildings such as Westminster Abbey and the Houses of Parliament are affected. Forty flying buttresses at Westminster Abbey, rebuilt only 90 years ago, need replacing already.

Many countries are trying to reduce pollution. In Britain, for example, the

3 How does acid rain affect ● trees? ● rivers and lakes? ● buildings?

4 Why is acid rain an international problem?

5 Study **D**.

   a Draw two divided bars similar to those for West Germany in **C**.

   b What differences are there between the sources of pollution in West Germany and Britain?

D Sources of sulphur dioxide and oxides of nitrogen air pollution in Britain

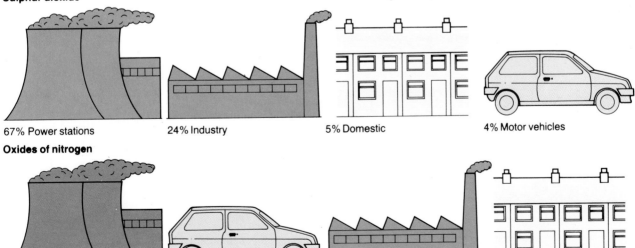

Sulphur dioxide

67% Power stations   24% Industry   5% Domestic   4% Motor vehicles

Oxides of nitrogen

46% Power stations   31% Motor vehicles   20% Industry   3% Domestic

2.13 Improving The Environment

When people interfere with the environment the result is often disaster. Every one of the people-made problems shown in **A** could be solved, but improving the environment requires careful planning, and extra money. Unfortunately, companies, local authorities and governments do not always wish to spend money on the environment.

---

1 Study **A** and write down why the following are environmental problems:
energy waste; river pollution; dirty beaches; traffic congestion
2 Copy and complete the following table:

Problem	Solution
energy waste	insulate houses with roof and wall insulation
river pollution	. . . . .

The Mediterranean

In 1975 all the countries bordering the Mediterranean Sea (except Albania) agreed to take action to protect the Mediterranean. **B** shows some of the environmental problems in the Mediterranean area. *Greenpeace*, the environmental campaign group, suggests the following ways in which people visiting the Mediterranean can help fight these problems:

1 Be careful not to start a fire
2 Do not destroy soil if you camp
3 Do not dump rubbish
4 Do not buy coral products; red coral is disappearing
5 Do not buy turtle products
6 Do not eat 'baby' fish
7 Do not go underwater fishing

The Camargue

The Camargue Regional Park is an example of conservation and protection. Its aim is to save and improve the environment. It is situated on the delta of the River Rhône (see Unit 2.9). The freshwater marshes and saltwater lagoons in the Camargue have

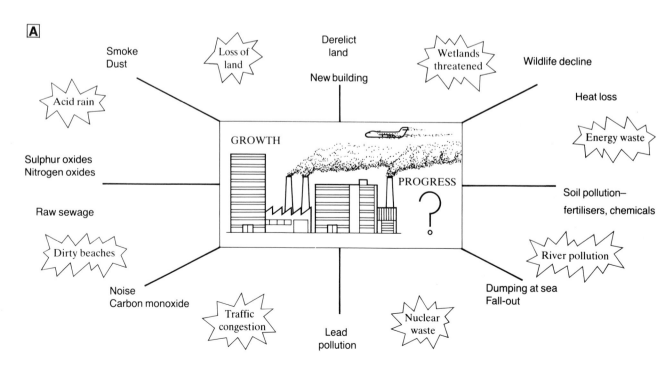

A

Smoke
Dust

Loss of land

Derelict land

New building

Wetlands threatened

Wildlife decline

Acid rain

Heat loss

Energy waste

Sulphur oxides
Nitrogen oxides

GROWTH

PROGRESS

?

Soil pollution—
fertilisers, chemicals

Raw sewage

Dirty beaches

River pollution

Noise
Carbon monoxide

Traffic congestion

Lead pollution

Nuclear waste

Dumping at sea
Fall-out

Skills	interpreting diagram, atlas work
Concepts	environmental damage, pollution, Nature Park
Issues	pollution control, conservation, role of individual

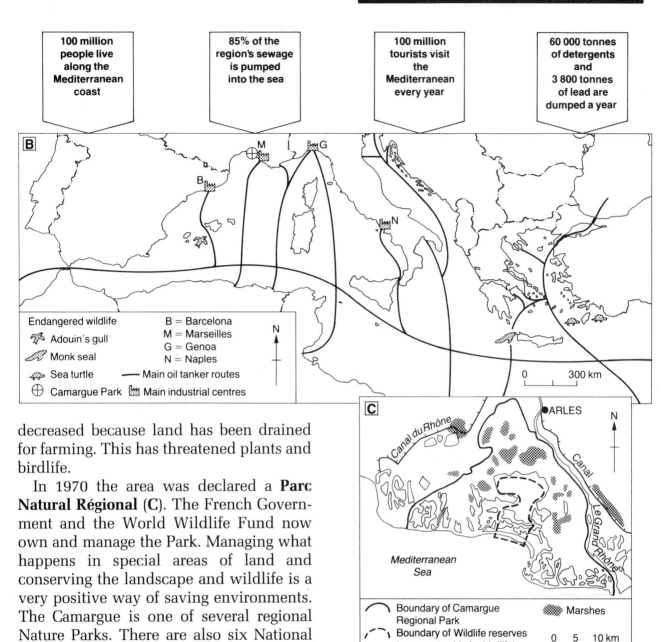

100 million people live along the Mediterranean coast

85% of the region's sewage is pumped into the sea

100 million tourists visit the Mediterranean every year

60 000 tonnes of detergents and 3 800 tonnes of lead are dumped a year

Endangered wildlife
- Adouin's gull
- Monk seal
- Sea turtle
- Camargue Park

B = Barcelona
M = Marseilles
G = Genoa
N = Naples
— Main oil tanker routes
Main industrial centres

0 300 km

C ARLES

Canal du Rhône
Canal
Le Grand Rhône

Mediterranean Sea

- Boundary of Camargue Regional Park
- Boundary of Wildlife reserves (areas set aside for birdlife)
- Marshes

0 5 10 km

decreased because land has been drained for farming. This has threatened plants and birdlife.

In 1970 the area was declared a **Parc Natural Régional (C)**. The French Government and the World Wildlife Fund now own and manage the Park. Managing what happens in special areas of land and conserving the landscape and wildlife is a very positive way of saving environments. The Camargue is one of several regional Nature Parks. There are also six National Parks in France.

---

3 How do you think the following pollute and upset the environment of the Mediterranean:
- oil tankers? ● industry? ● tourists?

4 Look at the advice given by Greenpeace. Why do you think points 5, 6 and 7 are important?

5 **a** Where is the Camargue?
b Why was the Camargue declared a Regional Nature Park?

6 Design a keyword plan for protecting and saving the Mediterranean (eg sewage treatment, wildlife reserves, limit dumping at sea . . .)

FURTHER WORK

- Think of an environmental problem not shown on **A**. Describe the problem and why it is a threat to the environment. How do you think the problem could be solved?

- Design a poster to explain the aims of a conservation project. (Use the example of the Camargue or another conservation project you know of.) The World Wildlife Fund (WWF) is a good source of information. It works to conserve places (habitats) now, so that they can be saved for the future.

59

3.1 Using resources

Western Europe has a high population density (see Unit 1.2). So many people living in a small area create a resources problem.

Western Europe does not have enough of the resources it needs:

- **oil** – Western Europe imports 50% of its oil
- **gas** – supplies of natural gas are running out
- **minerals** – iron ore, bauxite and copper are mainly imported
- **coal** – many of Western Europe's coal deposits are now exhausted. Coal is imported from Australia and South Africa

These are all **non-renewable** resources – once they run out they cannot be replaced.

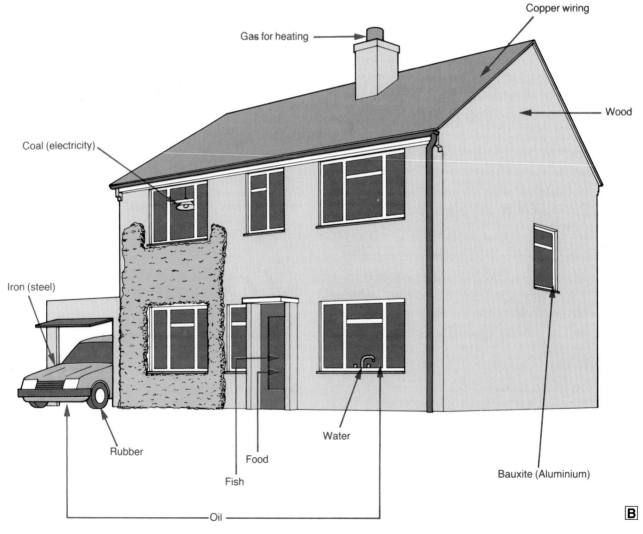

Skills	using diagram, designing poster
Concepts	renewable/non-renewable resources, recycling
Issues	management of waste, recycling resources

Western Europe also needs other, **renewable** resources:

● **natural rubber** – this has to be imported, or replaced by synthetic rubber
● **fresh water** – supplies are adequate, but water has to be used carefully
● **food** – Western Europe can produce more than enough food, but imports tropical products (eg bananas) and animal feed (eg maize)
● **fish** – stocks are declining
● **timber** – supplies of timber will only last if forests are replanted

Renewable resources need careful management. Fish must not be over-fished. Soil must not be polluted or used so much that it becomes infertile. Trees must be planted to replace those cut down. Care must be taken to cut down on wastage, to find substitute materials and to recycle products. **A** shows a 'bottle bank' used to collect glass for recycling.

Inputs and outputs

We can understand the resources problem better if we look at one household. In **B** the arrows represent some of the **inputs** – that is, resources going into the household system. **Outputs** (things going out of the system) include dirty water, sewage, heat (energy) loss, oil leaks, carbon monoxide, worn rubber, waste paper and leftover food.

You can see that there are a lot of resource inputs even for one household. Remember, there are over 330 millon people in Western Europe. On average, people are getting richer – they have bigger houses and cars, buy more washing machines and dish-washers ... in other words, they use up more resources.

How long can Western Europe go on using non-renewable resources at the present rate? Time is running out!

1 What resources do you use in your household? (Use **B** to help you.)

2 The diagram below shows how resources are used in a household system. Copy and complete the diagram for the following: paper; coal (electricity); water; a resource of your choice.

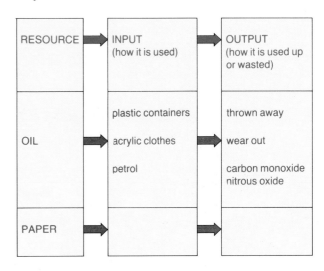

RESOURCE	INPUT (how it is used)	OUTPUT (how it is used up or wasted)
OIL	plastic containers acrylic clothes petrol	thrown away wear out carbon monoxide nitrous oxide
PAPER		

3 Choose two resources and say what your household could do to cut down on wastage (eg remembering to turn off the lights when you leave a room saves electricity; cycling instead of using the car saves fuel).

4 Which of the resources that you use could be **re-cycled** (ie used again)? Keep a week's diary of resources that you throw away and suggest how they could be re-used.

FURTHER WORK

● Design a poster warning people of the need for careful management of resources. Include an eye-catching heading, eg HOW LONG HAVE WE GOT? or *TIME IS RUNNING OUT!*

3.2 Building A Dam – Contours

A

B

Dams are built to hold back water (**A**). The **reservoir** behind a dam can be used for the generation of hydro-electricity, for irrigation or for water supply. Dams are used to control river flow – for example, to prevent flooding. A river can also be dammed to create a lake for recreation, for example fishing and water sports (**B**).

Damming a valley

Look at the contour shape of the valley in **C**. First study the diagrams on the left – 'before the dam'. Study the **cross-section** XY and the **long profile** ZP.

Then look at the diagrams 'after the dam' (on the right). A dam has been built across the valley. A reservoir has formed behind the dam, following the 100 metre contour line. Now study the map, cross-section and long profile 'after the dam'.

---

1 a Study **C**. How wide is the dam (Use the scale line on the map)?

 b How long is the reservoir at its longest point?

2 a Which way is the river flowing in the valley?

 b How deep is the reservoir at the deepest point of the cross-section?

Skills	sketching, measuring, cross-section
Concepts	dams and reservoirs, long profile, contours
Issues	environmental pressure

Valley map and cross-section

Before the dam

After the dam

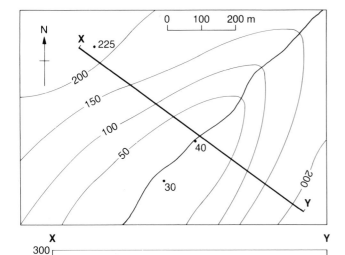

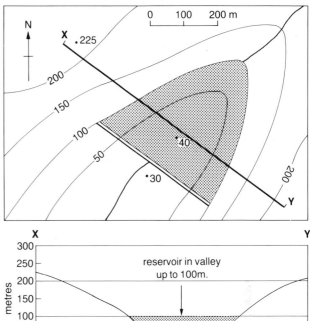

Valley map and long-profile

Before the dam

After the dam

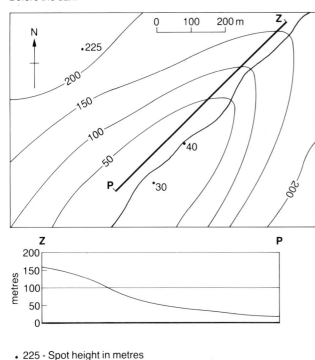

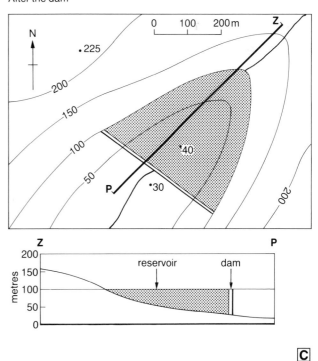

• 225 - Spot height in metres

C

3 The contour map does not show how the valley was used before the dam was built. Write down what could have been flooded when the reservoir was filled up. *Or* do the exercises in the *Activity pack* on choosing the reservoir site.

3.3 Hydro-Electricity

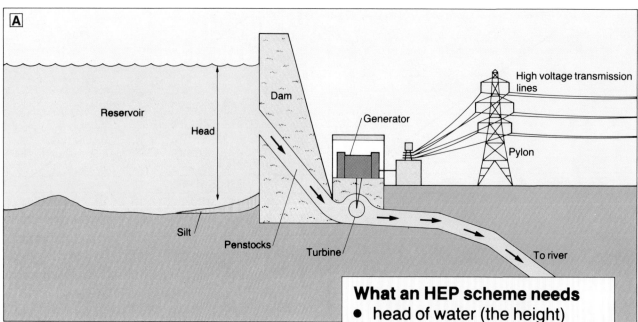

What an HEP scheme needs
- head of water (the height)
- large volume of water
- reliable rainfall
- no serious freezing
- impermeable rock (does not absorb water)
- a valley to dam
- a demand for electricity

Electricity is generated from rivers in the mountain areas of Europe. **Hydro-electric power** (HEP) stations use water which, unlike coal and oil, is a renewable resource (see Unit 3.1). This method is also pollution free, although reservoirs may destroy natural habitats and flood villages and farmland. **A** shows how a mountain river can be dammed to create a reservoir. Water flowing from the reservoir drives turbines which generate electricity.

Hydro-electricity schemes do have some disadvantages (**B**). They are expensive to build as they need large amounts of concrete and heavy generating equipment. They are often in high mountain areas and access is difficult.

1 What is HEP?
2 Study **A**. Without drawing a sketch describe how hydro-electricity is produced.
3 **a** What are the *advantages* of hydro-electricity?
  b What are the *disadvantages* of this type of power production?

4 Look at photograph **C**. Why is this a good site for an HEP station? Use the following words in your answer: deep valley, narrow valley, high rainfall, impermeable rock, large lake/reservoir.
5 Study the problem areas shown in **B**. Explain what each of the five problems might be.

HEP from the River Rhône

Hydro-electricity can also be generated on the lower courses of large rivers. The River Rhône in France now has 12 large HEP schemes (see Unit 2.9). The head of water (the drop in height) is not very large but the volume of water is great. Wide dams have

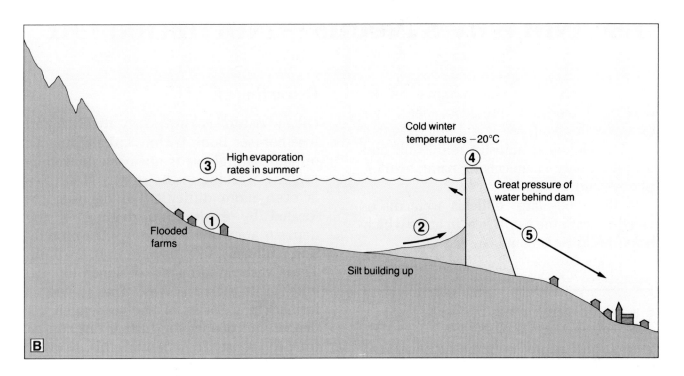

B

been built across the river. The new dams also
● control flooding;
● provide water for irrigation;
● provide new river crossings.
When a dam and reservoir have several uses they are called **multi-purpose**.

D is a **divided bar** graph which shows how important HEP is to some countries.

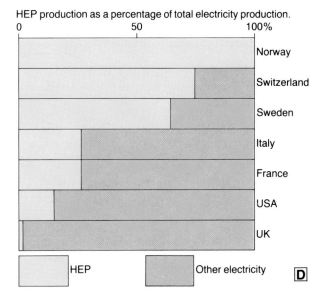

HEP production as a percentage of total electricity production.

| | HEP | | Other electricity | D |

C Hydro-electricity scheme at the Grimsel Pass, Switzerland

6 How do the hydro-electric power stations on the River Rhône differ from the power station in **A**?

7 List all the possible uses of a multi-purpose dam and reservoir.

8 Study **D**.
  a Why do you think all of Norway's electricity is hydro-electric?
  b What percentage of Switzerland's electricity is hydro-electric?
  c Why has the UK such a small percentage of HEP?

65

3.4 Norway's Needs – North Sea Oil

Oil is a very important resource to the developed world. It provides energy, for example diesel oil, petrol and home heating oil. The **chemical industry** refines oil and uses it to make chemicals for paint, textiles, plastics and fertilisers.

In the 1960s large **oilfields** were discovered beneath the North Sea, in areas which belonged to Britain and Norway (**A**).

---

1 a What industry refines oil?
  b What products is oil used for?
2 Why might people in Britain be pleased to discover oil in the North Sea?

Extracting oil

Oil and natural gas are nearly always found together (see Book 1, Unit 3.9). Drilling for oil and natural gas is always a dangerous process; working in the stormy North Sea is even more difficult. Oil deposits are located by **exploration drilling**. If the deposits are worth exploiting then **production platforms**, like the ones in **B**, are built. These vast platforms are designed to cope with giant storm waves. The seabed is formed in a series of shelves which get deeper and deeper the further out you go from the coast. To cope with this, Norway has built platforms for different depths of sea. **C** shows the Statfjord platform.

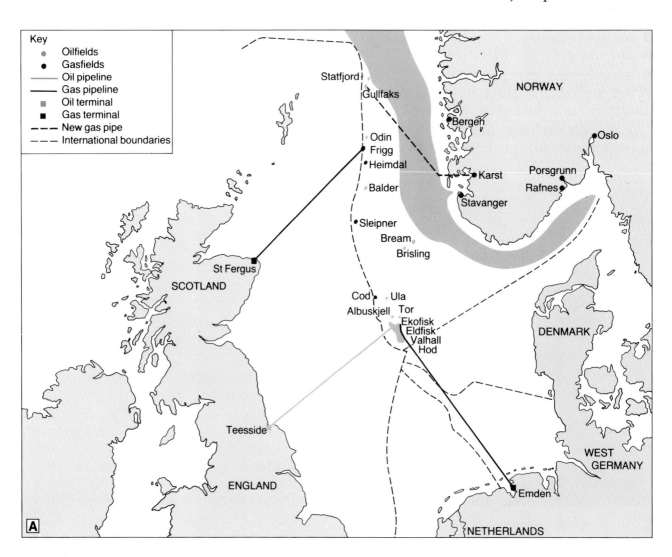

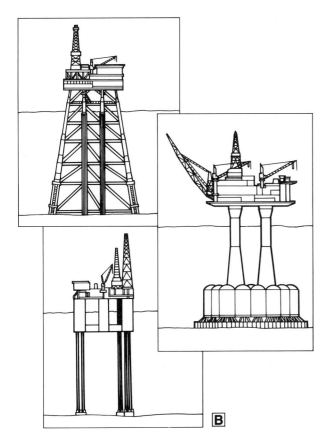

The oil and gas reach the platform at a temperature of 60°C. The two resources are separated and the oil is cooled before being transported to shore by tanker or pipeline. The gas is compressed and any water is removed before it is piped ashore.

---

3 Why is drilling for oil in the North Sea dangerous?

Skills	imaginative writing, interpreting diagram, mapwork
Concepts	chemical industry, oil fields, natural gas, refinery/processing
Issues	pollution, danger

---

4 How are oil deposits located?

5 Look at **B**.
  a Why are platforms built for different depths of sea?
  b What is the deepest sea they operate in?

6 How are the oil and gas transported ashore?

Where does the North Sea Oil go?

There is a 300-metre trench off the south west coast of Norway (see **A**), so all the oil from the central North Sea is piped to Teeside in England and Emden in West Germany. Gas is piped to St Fergus in Scotland. The Norwegians have now built a pipeline to take gas to their own country, at Karsto.

At the **refinery** the oil is broken down into many useful products, for example petrol and home heating oil. Natural gas is also processed to extract chemicals for industry. As well as being valuable resources in their own right, North Sea oil and gas help create jobs for many workers in Norway and in Britain (see **D**). By 1984 53% of Norway's earnings from exports came from oil-related products.

Jobs	Where	Doing what
1000	Rafnes	Petro-chemical works
8000	North Sea	Production platforms
22000	Norway	Platform and shipbuilding

D Jobs in the oil industry

FURTHER WORK

● This unit has described some of the **benefits** of the oil industry, eg energy for industry, wealth from exports; jobs. What **disadvantages** can you think of?

  Imagine oil is spilled into the North Sea. What damage could it do to the sea environment and the coastline?

● Look around your classroom. How many things are made from materials derived from oil?

3.5 The Power Crisis – On The Edge Of Darkness

On 26 April 1986 a leak of hydrogen gas caused an explosion and fire in reactor No.4 at Chernobyl nuclear power station in the USSR (**A**). Courageous firemen prevented the fire spreading to the other reactors, but at the cost of their own lives. The firemen were killed by **radiation** from the reactor which was exposed by the blast. A giant **radioactive** cloud drifted across Europe. At first the Russians did not reveal the disaster, but radioactivity was soon detected in Sweden and the Swedes broadcast the news.

In the days following the explosion, a radioactive cloud drifted back and forth across Europe, casting a shadow of death across the continent. It has been estimated that in Europe, outside the USSR, up to 2000 people will die from cancers due to radioactive fallout. In the USSR the final figure could be much higher. Only two people were killed in the fire, but by September 33 people, including the brave firemen, had died from radiation sickness.

135 000 people had to be evacuated from the area surrounding Chernobyl. Food and milk had to be destroyed. As far away as Britain, sheep which had grazed contaminated grass in the mountains of Wales and the Lake District had to be slaughtered because their meat would contain high levels of radioactivity. There was a ban on moving and selling sheep. The ban lasted for several years after the disaster and cost many farmers a lot of money. Poland claimed that it had lost over £6 million because 40 000 tourists had cancelled holidays for fear of the radiation. The way of life of the Lapps in Scandinavia was devastated by the build up of radioactivity in their reindeer which grazed contaminated moss and lichen. The survival of the Lapps depends upon their reindeer. None of the reindeer meat could be safely eaten in 1986.

A The Chernobyl power station after the explosion.

Chernobyl cast a cloud over the future of nuclear power. Public opposition to nuclear power increased in many countries. The great dreams of a future based entirely on cheap, clean nuclear electricity were shattered. **B** shows percentages of nuclear electricity for some Western European countries.

Country	Percentage		
	1980	**1985**	**1995** (Projected)
Britain	13	19	26
France	24	65	75
West Germany	11	31	33
Italy	1	4	13
Spain	5	22	30
Belgium	23	63	66
Netherlands	6	6	13
Sweden	22	42	47
Switzerland	28	40	36

B Nuclear power as a percentage of total electricity production

Sources of energy

After what happened at Chernobyl, many people think that nuclear power is too

Skills	interpreting statistics
Concepts	radioactivity, nuclear power, alternative energy
Issues	nuclear power

dangerous to be a good source of energy. Oil is too precious to burn for power generation. Coal-fired power stations cause acid rain. What are the alternatives for electricity generation? Hydro-electricity (Unit 3.3)? There are few undeveloped sites left in Europe. **C** shows some other ideas for alternative energy.

Wind
Giant windmills called wind turbines could supply one tenth of Britain's electricity by 2000. 100 metre high wind turbines can generate up to 5 MW. They would have to be built in groups of several hundred to generate the same power as one coal-fired power station.

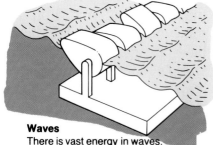

Waves
There is vast energy in waves. Norway has an experimental wave power station which produces 1 MW of electricity. Much work needs to be done to produce large scale economic wave power stations.

Tides
A tidal power station on the estuary of the River Rance in France generates 240MW of electricity. 24 turbines are housed in a hollow dam over 700 metres wide. There are plans for tidal power stations on the estuaries of the Rivers Severn and Mersey in Britain.

Geothermal
Some areas have natural hot water beneath them which can be used to provide heating and to help produce steam to generate electricity. There are geothermal power stations in Iceland, New Zealand, Italy and the USA. The 'hot dry rock' method is a form of 'renewable' geothermal energy which involves pumping water down a deep borehole into hot granite. The heated water rises up a second shaft to the surface where its energy can be used to drive turbines. Tests have been carried out at Camborne in Cornwall.

Solar
Solar power can be used to generate electricity, but it is very expensive. A 10 MW power station exists in California, USA and there are plans for a 100 MW station. Solar panels heating water in the roof could replace electricity as a heat source in the summer, saving about 2500 MW of power in Britain alone. **C**

---

1 a Where is Chernobyl?
  b What happened there in April 1986?

2 What were the effects of the Chernobyl disaster?

3 a Which country in table **B** had the highest percentage of electricity generated by nuclear power stations (i) in 1980 and (ii) in 1985?
  b Which is the only country that planned to reduce the percentage contribution of nuclear power by 1995?
  c What effect do you think the Chernobyl disaster might have on the actual contribution of nuclear power by 1995?

4 Look at **C**.
  a What are wind turbines?
  b What percentage of Britain's electricity could be generated by wind power by the year 2000?

5 a Which countries have: (i) wave; (ii) tidal; (iii) geothermal power stations?
  b What is the 'hot dry rock' form of geo-thermal energy?

6 The forms of energy shown in **C** are all 'renewable' forms of energy. What do you think this means? Can you think of some 'non-renewable' forms of energy?

3.6 Resources Run Out

Gold rush! Diamonds! Resources attract people. They do not have to be precious minerals. Coal and iron ore sound less romantic than gold and diamonds, but they have attracted more people. Many of the world's biggest industrial regions have grown up because of resources of coal and iron ore. In Britain, industrial areas in the West Midlands, South Wales, Central Scotland and North East England all developed around coal and iron ore fields. During the Industrial Revolution in the 19th century people flooded into such areas seeking work in the mines and factories which used the resources.

But what happens when the resources run out? You have probably heard of 'ghost towns'. These are abandoned towns which grew quickly to mine gold and died just as quickly when the gold ran out. Are there coal ghost towns? In County Durham, in North East England, pit villages built around some coal mines were demolished once the mine closed. The populations of whole regions are threatened when the resources run out.

Lorraine is a region in eastern France. During the 19th century it grew to become one of France's largest industrial regions (**A**). The growth was based on mineral resources: coal, iron ore and rock salt. These resources supported vast mining, steel and chemical industries.

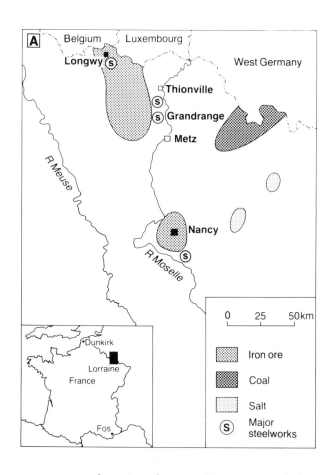

---

1 Where is Lorraine?
2 a Which resources were the cause of Lorraine's industrial growth?
  b What industries developed in Lorraine based on those resources?

The first large iron ore mines opened in the 1880s. Iron and steelworks were built close to the mines around Longwy, Thionville and Metz. In 1970 a large new steelworks was opened at Gandrange (**B**). Many of the smaller works were then closed because they were losing money. By 1980 even Lorraine's larger steelworks were losing money. Lorraine's iron ore and coal had not run out, but they were much more

Skills	line graph
Concepts	unemployment, industrial decline, emigration
Issues	unemployment

expensive than the resources imported from other countries. This was because Lorraine's ore was low grade and expensive to mine. It was cheaper to import iron ore 7000 km from Brazil than to use Lorraine's own ore.

Because most resources are imported the best location for a steelworks in the 1980s is on the coast. There are two huge new works, at Dunkirk on the north coast (**C**) and at Fos near Marseilles. These produce over half of France's steel. Lorraine's days as a major steel producer are numbered. Only the Gandrange works may have a long-term future.

	1960	1987
Iron ore production (million tonnes):	63	17
Coal production (million tonnes):	15	9
Steel production (million tonnes):	12	5
Iron ore miners (thousands):	33	8
Coal miners (thousands):	27	14
Steel workers (thousands):	105	38

D Industrial decline in Lorraine – the statistics

D shows how Lorraine's traditional heavy industries have declined since 1960. Many people lost their jobs and were forced to seek work elsewhere in France. The population of Lorraine fell by 12 000 between 1975 and 1982. The French government has provided grants and loans to attract new industry into the region. Car components, food processing, engineering, computers and clothing industries have been set up at Thionville, Metz, Nancy and Longwy. But many more new jobs will have to be created if Lorraine is to fully recover.

3 Study **D**:
   a How many workers were employed in Lorraine's mining and steel industries in (i) 1960 and (ii) 1987?
   b What has caused the changes?
4 Draw a line graph showing the population of Lorraine between 1962 and 1982; using the following statistics:

	1954	1962	1968	1975	1982
Population (millions):	1.96	2.19	2.26	2.33	2.21

5 Study photographs **B** and **C**. Describe the scene in each photograph and state what advantages Dunkirk has over Gandrange as the location for a modern steelworks (think about space, transport).
6 **a** What can be done to help regions of high unemployment like Lorraine?
   b What advantages and disadvantages do you think Lorraine has as a location for modern industry such as electronics? (Hint: think of Lorraine's position within Western Europe and the importance of the EC).

3.7 Preserving Antarctica

A land of ice, penguins and seals. . . an area of cool summers and bitterly cold winters. . . In December there is permanent light; in June there is total darkness. This is **Antarctica**, a continent which covers one-tenth of the Earth's land surface.

Antarctica is the world's last 'wilderness'. It was the last continent to be discovered and mapped. In 1911 the Norwegian explorer, Roald Amundsen, became the first person to get to the South Pole.

16 nations now control activities in Antarctica. They have agreed to keep the area **demilitarised** (with no missiles, weapons or military bases); **nuclear-free** (with no nuclear power stations or bomb stores) and left for **research** (scientific study and exploration). Seven nations claim sectors of the continent (**A**).

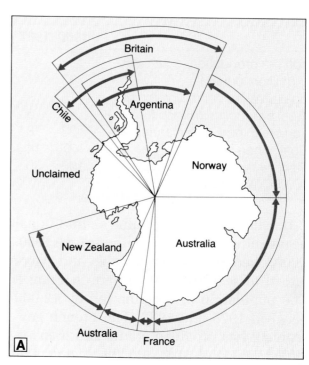

A

Skills	atlas work, cloze text, designing poster
Concepts	environmental threat, food chain
Issues	environmental pressure, conservation

The environment

Antarctica is the largest wildlife sanctuary on Earth. It is rich in animal and sea-life. Its glaciers and icefields contain 90% of the world's fresh water. The average depth of the ice is 2000 metres. If all this ice melted, the world's sea level would rise by 90 metres! The land area of Antarctica is surrounded by pack-ice, which breaks up slowly in the short summer.

The Antarctic area is home to colonies of penguins, seals and whales. Many species of fish live in large shoals, and there are huge swarms of **krill**. Each part of this **ecosystem** is linked with another; what happens to one species of animal affects another. This ecosystem can also be presented as a **food chain** (**B**).

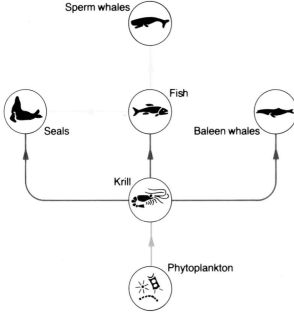

B

Krill are shrimp-like crustaceans which grow to about five cm. They are the food for many other species of marine life.

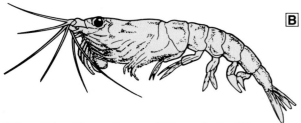

This diagram shows a simple food chain. Each line shows one species eating another.

So far, people have not upset Antarctica, but its security may be threatened:

- There could be territorial disputes among the seven nations which claim territory in Antarctica.
- Many other countries may want a stake in the continent.
- Oil could be exploited, leading to harmful development and pollution.
- Minerals may be mined.
- There is already overfishing in some areas.

---

1 Look at **A** and find an atlas map showing Antarctica. Copy and complete the following passage – fill in the gaps from the **word-bank** below.

The land of Antarctica is a . . . . . Surrounding the land is the . . . . . Ocean. The seas near the land are covered by . . . . The line of . . . . . at 66½° South is called the . . . . The first explorer to reach this point was the Norwegian . . . . . Today . . . . . nations claim parts of the land. Only . . . . . live in the area and there is no farming, no industry and no towns.

Amundsen, scientists, Antarctic, Antarctic Circle, South Pole, continent, seven, pack-ice, latitude

2 Study **B**.
  a What is krill?
  b Why will Baleen whales die if krill is overfished?
  c Why will the Sperm Whales and seals not survive if krill is overfished?

3 Discuss what would happen to Antarctica's ecosystem if krill were killed off by sea pollution.

4 You now know the threats to Antarctica. What do you suggest should be done to save the continent and surrounding seas?

FURTHER WORK

- Design a 'Save Antarctica' poster which shows features of the environment and marine life.
- Do the exercise on Antarctica in the *Activity book*.

3.8 Food for the Future: Science and Farming

The world's population will double in the next half century. This growing population (80 million extra people per year) will need feeding. How can farmers meet this demand?

The present

The biggest progress in agriculture so far has been in the developed world, where industrialised nations can afford the cost of research and development, and farmers can afford to buy new technology. Farmers have become more efficient: using more sophisticated machinery (**A**); trying new varieties of plants (**B**) and new ways of growing, harvesting and storing crops; developing scientific methods of rearing animals (**C**) and finding new uses for farm produce.

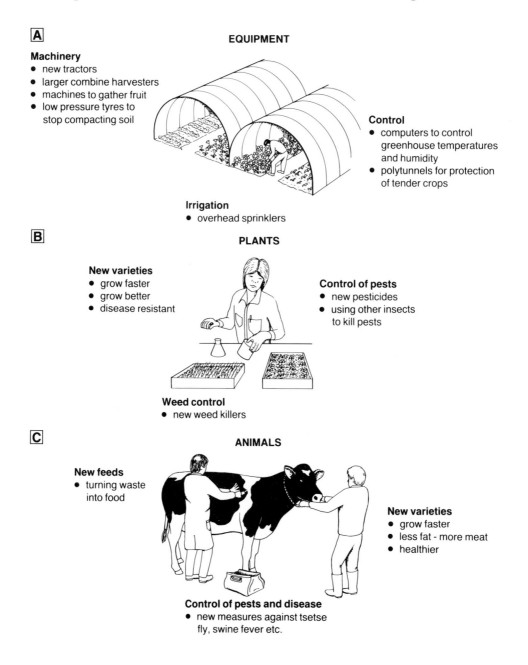

A EQUIPMENT

Machinery
- new tractors
- larger combine harvesters
- machines to gather fruit
- low pressure tyres to stop compacting soil

Control
- computers to control greenhouse temperatures and humidity
- polytunnels for protection of tender crops

Irrigation
- overhead sprinklers

B PLANTS

New varieties
- grow faster
- grow better
- disease resistant

Control of pests
- new pesticides
- using other insects to kill pests

Weed control
- new weed killers

C ANIMALS

New feeds
- turning waste into food

New varieties
- grow faster
- less fat - more meat
- healthier

Control of pests and disease
- new measures against tsetse fly, swine fever etc.

Skills	keyword plan, imaginative drawing
Concepts	problems in farming, biotechnology, organic farming, natural predators
Issues	chemicals v. organic farming, healthy eating, waste

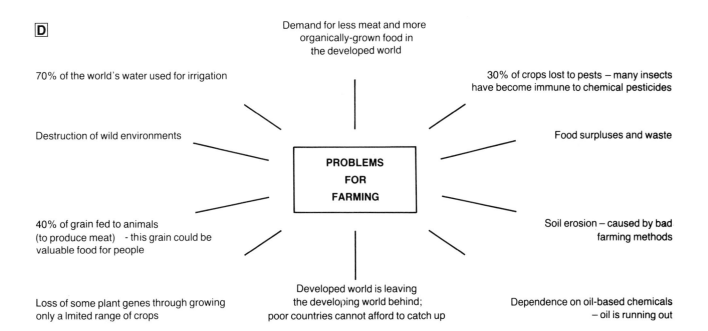

D

Demand for less meat and more organically-grown food in the developed world

70% of the world's water used for irrigation

30% of crops lost to pests – many insects have become immune to chemical pesticides

Destruction of wild environments

Food surpluses and waste

PROBLEMS FOR FARMING

40% of grain fed to animals (to produce meat) - this grain could be valuable food for people

Soil erosion – caused by bad farming methods

Loss of some plant genes through growing only a lmited range of crops

Developed world is leaving the developing world behind; poor countries cannot afford to catch up

Dependence on oil-based chemicals – oil is running out

However, all this progress is costly – in money, in scarce resources and in possible damage to the environment (see **D**)

The solutions

Farmers have always tried to change their land to suit their crops. Now, science is helping them to change the crop to suit the land. **Biotechnology** is helping to produce plants that can grow in drier conditions without much attention from the farmer. Farmers can also grow particular plants to put back goodness into the soil: alfalfa, for example, replaces nitrogen in the soil through its roots.

New methods of working the land have been developed, which do not compact the soil. Plastic tunnels are used to protect crops (see **A**).

Over the past ten years many people in the developed world have become concerned about the food we eat. 'Healthy eating' campaigns have encouraged people to adopt a healthier lifestyle. Consumers are worried about the use of chemicals on food crops and the demand for organic foods is increasing.

Organic farming methods do not involve toxic chemicals, fertilizers and pesticides.

The farmer feeds the soil, not the plant. Organic material (that is, anything that lives, such as plants) is used to improve the structure and fertility of the soil. Organic farmers use **natural predators** to control pests – for example, ladybird beetles are released in greenhouses to feed on the aphids which damage crops.

1 What percentage of crops are destroyed by pests?
2 Give three examples of developments in **a** plants; **b** animals and **c** equipment.
3 Why is bio-control being developed? Give an example.
4 How much of the world's water is used for irrigation?
5 What is organic farming?

FURTHER WORK

● Design a poster to show some of the ways in which you think future farming will develop.

3.9 Resources under Threat – Timber and Fish

Western Europe was once largely covered by forest. Today, only small areas of this great forest remain. People have felled the trees in their millions. This process began thousands of years ago when people first began to use wood for fuel and to build shelters. Later vast areas were cleared to provide land for farming.

Deforestation has stripped the landscape of its trees and caused problems in several areas (**A**). Remove the trees and you remove protection for the soil. Tree roots bind the soil together and the leaves and branches break the force of the rain. Once the trees are gone, rain washes the soil away (**soil erosion**). This is especially bad on slopes. Soil is washed down into rivers, it chokes the channels and flooding results.

One area badly hit by soil erosion is Southern Italy. Many of the slopes are completely bare of soil. Farming this landscape is almost impossible. Over time the flooding caused by soil erosion formed marshes which became breeding grounds for mosquitoes carrying malaria. People left the river valleys to live in villages on the hilltops, away from the disease.

Today the forestry industry is carefully managed in most European nations (**B**). When trees are felled, new ones are planted to replace them. Foresters realise that a reasonable amount of timber can be extracted without any problem. Timber is a **renewable resource** which will always be available as long as it is not **over-exploited**.

Fishing

Fish is another renewable resource (**C**). Europe's fish stocks have been badly hit by **over-fishing**. The North Sea herring provides a good example. Such large numbers of herring were caught, including very young fish, that some experts thought the herring would become **extinct**. The EC imposed a ban on herring fishing in 1977. The ban was lifted in 1983 when the herring stocks were said to have recovered.

A Before

After

Skills	anagrams, describing photograph
Concepts	deforestation, soil erosion, renewable resource, over-fishing
Issues	people's effect on the environment

1983 also marked the introduction of a **Common Fisheries Policy** (CFP) by the EC. The CFP sets **quotas** (a share of the total fish allowed to be caught) for each nation's fishing fleets. It also sets a minimum **mesh size** for fishing nets.

---

1 a Give three reasons why people have cleared much of Europe's forests.
 b List all the things in your classroom made of wood.

2 Study **A** carefully.
 a Write a sentence about both the drawings.
 b Unscramble these anagrams to provide three words which describe what is happening in **A**:

 I OIL ROSE SON   TEN FOOTED AIRS

3 What problems have been caused by soil erosion in Southern Italy?

4 a Describe the scene in photograph **B**.
 b How can forests be useful when they are 'just standing there'? (Hints: leisure, wildlife)

5 'Timber is a renewable resource.' What does this mean?

6 a In what way are fish an example of a renewable resource?
 b What does 'over-fishing' mean?
 c What do the letters CFP stand for?
 d Why do you think it was necessary for the CFP to set a minimum mesh size for fishing nets?

3.10 Health and Education

What are people's basic needs? A group of pupils were asked this question. You can see some of their ideas in the keyword plan, **A**.

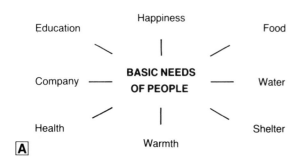

Education   Happiness   Food

Company — **BASIC NEEDS OF PEOPLE** — Water

Health   Warmth   Shelter

A

The human body is a well-designed machine, capable of a wide range of physical movements. The human brain can store a vast amount of information and perform many complicated calculations very quickly. **B** shows what we need to keep the body in good condition.

Health

People in Europe and the developed world expect their babies to live and grow up strong and healthy. They expect to get good medical care when they are ill. Health care is Britain's biggest service industry, employing over a million people. We take good health and health care for granted (**C**).

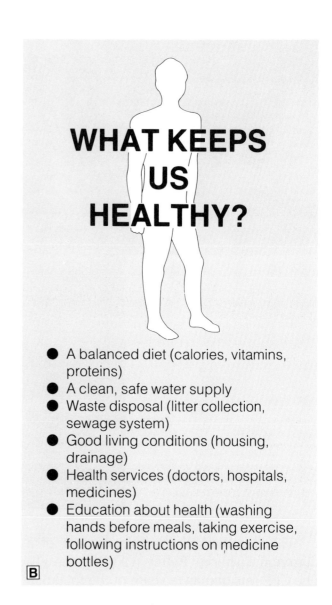

WHAT KEEPS US HEALTHY?

- A balanced diet (calories, vitamins, proteins)
- A clean, safe water supply
- Waste disposal (litter collection, sewage system)
- Good living conditions (housing, drainage)
- Health services (doctors, hospitals, medicines)
- Education about health (washing hands before meals, taking exercise, following instructions on medicine bottles)

B

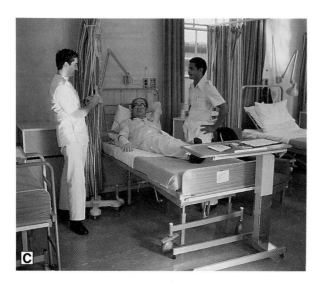

C

People in Britain have the National Health Service (NHS) and private medical care schemes such as BUPA. If someone cannot afford to pay for medical services the NHS will take care of them. This public service is paid for by taxes which the Government collects.

Such good medical care is only available to people in the developed world. And even in developed countries, there are health problems. Health has become a world issue (see **D**).

Disease does not recognise countries' frontiers. Improved travel means that people can cross the world in hours; they

Skills	keyword plan, poster work, empathy, research
Concepts	health education, national/private health care, basic needs
Issues	world health

1 Average life expectancy at birth is 72 years in the developed world but only 54 years in the developing world.

2 Infant mortality in the developed world is 18 deaths for every 1000 babies in their first year. In the developing world 101 per 1000 babies die in their first year.

3 The developed world has three times as many people aged 64 and over as the developing world.

4 20% of all deaths in the developed world are due to cancer. In some developed countries 50% of deaths are due to heart-related diseases.

D Reasons for world concern about health

can contract and spread disease quickly. AIDS is one example of a disease that has spread rapidly as a result of people's increased mobility.

The World Health Organisation (WHO) was set up in 1948 to help countries develop health programmes and to take action to check the spread of disease, for example malaria, leprosy and yellow fever. Already smallpox has been successfully eradicated.

1 **a** Make a copy of keyword plan **A**.
  b In a group, discuss what else you would add to the list and give reasons for your ideas.
2 **a** How is the human body a 'well-made machine'?
  b What do people need to keep healthy?
3 **a** What do the initials NHS and WHO stand for?
  b Why is health an issue of world concern?

Education

Education teaches you to read and write and gives you the knowledge and skills you need to survive and develop. Education has high priority in all countries.

Many governments provide a **formal education**; children go to school (**E**) and may then go on to higher education, for example university.

If parents have enough money they may send their children to **private** schools. Otherwise the children go to **state** schools. The developed world needs a highly trained workforce and so requires a good education system. In the highly industrialised nation of Japan, 90% of 15-year-olds stay on at school. Japanese universities produce ten times more engineers than those in Britain.

E

4 **a** What is education?
  b Why is education important to governments?
  c What is formal education?
5 Copy out your timetable of lessons for the week.
  a How many different subjects do you study?
  b How does each subject help you develop as a person? What skills do they teach you?

FURTHER WORK

● How can education keep people healthy? Design a poster telling people how to look after their health. (**B** may help you.)

3.11 The Future of the Rich World

Some days the TV and newspapers are full of doom and gloom. There seem to be so many problems facing the world. Even the rich world seems full of problems:

Economic problems
rising prices
falling industrial output
rising unemployment

Environmental problems
acid rain
sea pollution
inner-city dereliction

Resource problems
oil running out
over-fishing
forests dying

Social problems
rising crime
drug problems
racism

In spite of all this, the rich world continues to prosper. New technology helps solve some problems, new methods are introduced. The future of the rich world often seems bright.

The snakes and ladders game

For two players

1 You both live in the rich world.

2 You need two counters and a die.

3 Throw the die to decide who is RED and who is BLUE. The person throwing the highest number is RED. (If you both throw the same then try again.)

4 If you are RED you only follow the red snakes and ladders. If you are BLUE you only follow the blue ones.

5 The first one to reach Number 48 is the winner. The one who gets on best in the rich world!

1 What happened to you in the game? Were you very fortunate or were you unlucky?
If you were RED you probably won because red had better chances than blue. The red snakes are shorter!

Life chances

Getting on well in the rich world is often a matter of chance like a game of snakes and ladders.

Right at the start of the 'game', some people seem to have little chance of getting on well. They are **disadvantaged**. These groups include:
● immigrant groups
● ethnic minority people
● one-parent families
● rural people
● inner-city people
● the old
● the unemployed

2 What other groups can you add to the list of disadvantaged people?

3 Now list the groups of people who you think *do* get on well in the rich world, eg the well educated.

FURTHER WORK

● Make up your own Snakes and Ladders game. You can use the outline in the *Activity pack*.

When you make up your game use red for the person getting on well. Use blue for the person with less chance of doing well. **Make the red snakes shorter than the blue snakes and the red ladders longer than the blue ones.** You could set up the game for a pair of people like the following:

rural and urban	old and young
black and white	rich and poor

Skills	co-operation, empathy, playing/designing game
Concepts	future, life chances, disadvantage
Issues	inequality, role of individual

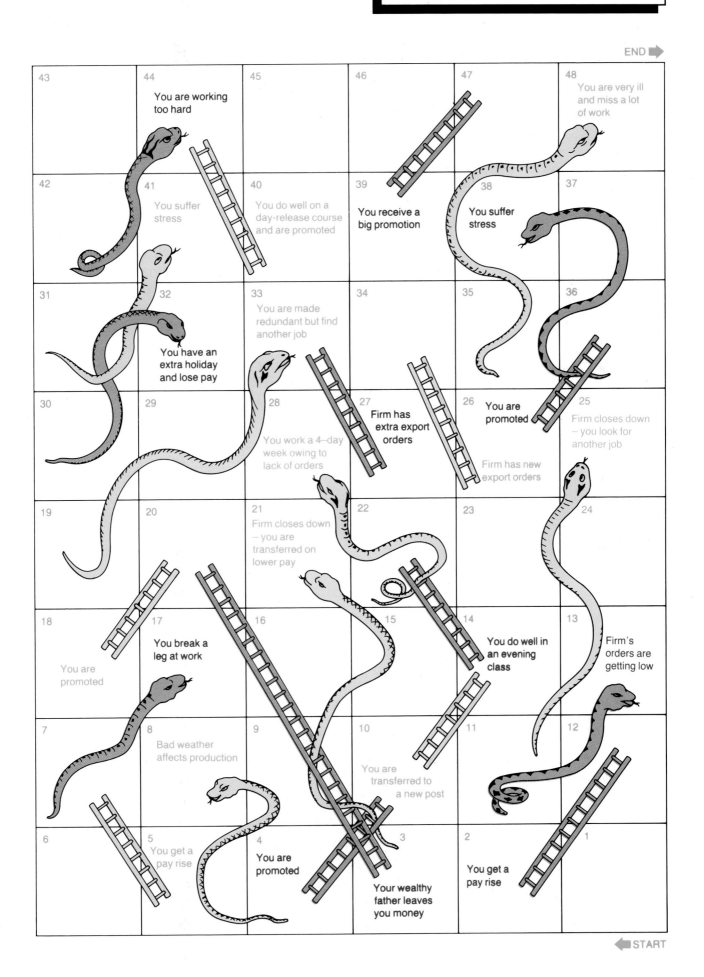

END ➡

43	44 You are working too hard	45	46	47	48 You are very ill and miss a lot of work
42	41 You suffer stress	40 You do well on a day-release course and are promoted	39 You receive a big promotion	38 You suffer stress	37
31	32 You have an extra holiday and lose pay	33 You are made redundant but find another job	34	35	36
30	29	28 You work a 4-day week owing to lack of orders	27 Firm has extra export orders	26 You are promoted / Firm has new export orders	25 Firm closes down – you look for another job
19	20	21 Firm closes down – you are transferred on lower pay	22	23	24
18 You are promoted	17 You break a leg at work	16	15	14 You do well in an evening class	13 Firm's orders are getting low
7	8 Bad weather affects production	9	10 You are transferred to a new post	11	12
6	5 You get a pay rise	4 You are promoted	3 Your wealthy father leaves you money	2 You get a pay rise	1

⬅ START

81

4.1 Farming Systems – Capitalism and Communism

The USA and the USSR both have populations of over 200 million people. Both countries have the right conditions for growing wheat (**A**) and large areas of farmland are used for **extensive wheat production**. Machinery is used to farm large areas – up to 1000 hectares per farm in the USA and 50 000 hectares in the USSR. According to world trade figures, the USA exports 34% and the USSR 13% of all wheat exports.

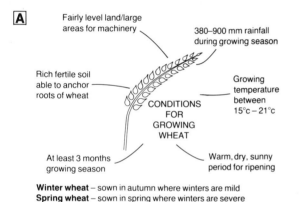

A

Fairly level land/large areas for machinery

380–900 mm rainfall during growing season

Rich fertile soil able to anchor roots of wheat

Growing temperature between 15°C – 21°C

CONDITIONS FOR GROWING WHEAT

At least 3 months growing season

Warm, dry, sunny period for ripening

Winter wheat – sown in autumn where winters are mild
Spring wheat – sown in spring where winters are severe

Steppelands of the USSR – communist system

B shows the main wheat-growing regions in the USSR. Soviet farming is organised by the state. The government owns and controls the use of land. The **sovkhoz** are huge state farms run by government officials. These farms are highly **mechanised**. Farmworkers are paid a set wage and live in small towns with their own schools, hospitals and food processing factories.

Since 1982 a **Food Programme** has operated in the USSR, to improve the diet of the people. It is planned to achieve this by increasing grain yields (amount of grain produced from each hectare) by, for example, using extra fertiliser. Part of the Food Programme includes improving the social and living conditions of farmworkers by building modern flats for them and their families to live in. Clubs, sports stadiums,

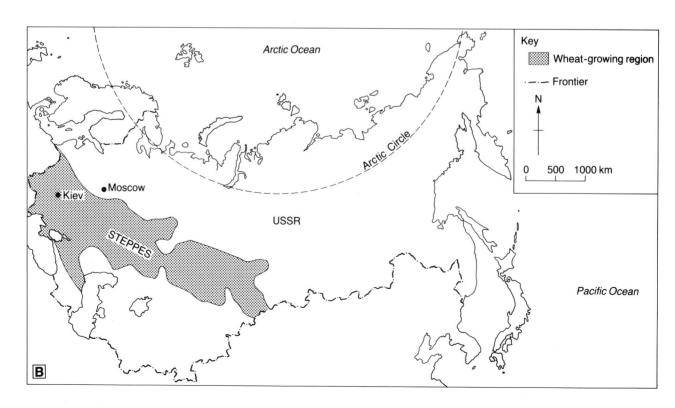

B

Key

Wheat-growing region

Frontier

N

0 500 1000 km

Arctic Ocean

Arctic Circle

Moscow

Kiev

USSR

STEPPES

Pacific Ocean

Skills	keyword plan, groupwork, empathy, decision making
Concepts	extensive wheat production, mechanisation, capitalist/communist farming
Issues	capitalism/communism

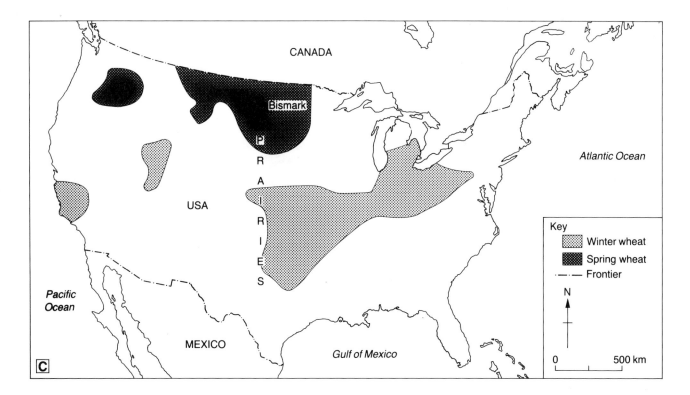

public baths, creches and laundries have also been provided.

In recent years there have been poor harvests, due in part to pollution (acid rain) and in part to poor weather (bad storms at harvest time). The Soviet Government has been forced to import wheat from the USA and the EC.

The Prairies of the USA – capitalist system

C shows the main wheat-growing regions of the USA. In the USA wheat farmers usually own their land and make their own decisions about how to use it. To pay for improvements, they borrow money from banks and pay interest on the loan. Although farmers can organise themselves (unlike Soviet farmers) they have the pressure of repaying loans. Failure to do so could mean losing their farm and livelihood.

Profit from the farm goes to the farmer, whereas in the USSR it goes to the state and the people. Private ownership means that farms are smaller in the USA. The government does help the farmer with grants towards improving production and storage

of wheat, but mostly American farmers have to pay for what they need: land, seed, labour, buildings, hire of machinery at harvest time and so on. As in the USSR, weather can be a problem for American wheat farmers, with lack of rainfall and long, severe winters.

---

1 What is meant by 'extensive wheat production'?
2 Use **A** to make a keyword plan for the conditions needed to grow wheat.
3 Look at **B** and **C**. Where is wheat grown in the USSR and USA?

FURTHER WORK

● In pairs, imagine one of you is a Soviet farmworker and the other is an American farmer. Discuss each other's way of organising farming. Compare the advantages and disadvantages of each system.
Now collect other ideas from people in your group.
● Given a choice, which system would you prefer to work in?
● Give reasons for your answer.

4.2 Food for Europe – The Netherlands

Edam cheese, tulips, ham, *Heineken* lager – these are all products of Dutch farming which are exported all over Europe. Dutch farming is among the most efficient in the world. There are over a quarter of a million people working on Dutch farms, that is 5% of the total workforce. A quarter of all Dutch exports are food or food products. Dutch farming is big business. **A** shows land use in the Netherlands.

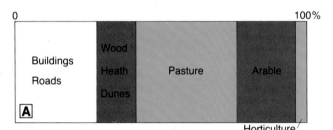

Dutch farming is geared towards exports. In fact, the Netherlands is second only to the USA as the world's leading exporter of farm produce! The Netherlands is the world's leading exporter of cheese, butter, condensed milk and milk powder. How is it so successful?

Specialisation
The Dutch have **specialised** on dairy farming, pigs, poultry, market gardening and flowers. They have concentrated their efforts on this small range of farming and become world leaders.

Co-operative associations
Dutch farmers have joined together to form over a thousand **co-operative associations**. These operate in every part of the farming sector, allowing farmers to buy seeds, fertiliser and feedstuffs in bulk, and so more cheaply. Members of co-operatives share expensive machinery and help each other with production, processing and marketing. The co-operatives set very high standards which all the farmers who belong to the co-operative must meet.

Science and technology
Dutch farmers make much use of new methods, including machines and computers, to boost production. Agricultural research and a well-organised agricultural advisory service are very important in the Netherlands. Much use is made of greenhouses like those in **B** with artificially-controlled environments.

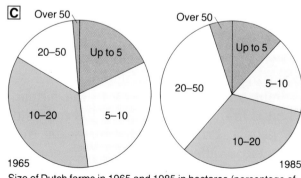

Size of Dutch farms in 1965 and 1985 in hectares (percentage of total farms)

Increasing size of holdings

Many small Dutch farms have closed (**C**). The average size of Dutch farms has increased from 17 hectares in 1970 to 22 hectares in 1986. At the same time the number of farmworkers has fallen from 340 000 to 270 000. Table **D** shows changes in Dutch agriculture between 1965 and 1985.

Dutch horticulture

Dutch bulbs have been famous for a long time. More recently there has been a rapid increase in cut flowers and pot plants (**E**).

1 Study **A**.

What percentage of the Dutch land area is made up of (i) arable land (ii) pasture (iii) horticulture?

2 Study **C**.

  a Which farm sizes decreased as a percentage of total farms?

  b Which farm sizes increased as a percentage of total farms?

  c Why do you think these changes happened?

3 **a** Draw a bar graph to show the numbers of cattle shown in **D**.

  b How do you explain the increase in livestock numbers, but decrease in the number of farms and farmworkers?

4 Study **E**

  a Where are the main areas growing flowers and pot plants in the Netherlands?

  b How can the Dutch farmers grow flowers and pot plants all year round?

  c How are the pot plants and flowers marketed?

	1965	1970	1975	1980	1985
Milk production (million tonnes)	7.0	8.5	10.3	11.8	13.2
Cheese (thousand tonnes)	219	284	375	441	530
Cattle (millions)	3.8	4.3	4.9	5.0	5.5
Pigs (millions)	3.5	5.7	7.3	10.1	11.1
Poultry (millions)	45.6	56.2	68.1	81.2	82.2
Workforce (thousands)	404	339	311	294	271
Number of farms (thousands)	153	126	109	96	88

D Changes in Dutch Agriculture 1965 – 1985

E

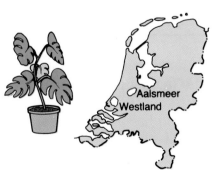

Cut flowers and pot plants are grown in vast greenhouses in the Westland area and at Aalsmeer.

The greenhouses have a completely artificial environment which gives the heat and light that the Dutch climate cannot provide. This allows the Dutch to grow plants all year round. The Netherlands has ample supplies of cheap, clean natural gas which is used to heat the greenhouses. Pot plants are placed on heated tables. Atomisers feed carbon dioxide into the atmosphere. Each pot receives water and fertiliser through computer-controlled drip-feed systems.

The flowers and pot plants are sold at the Aalsmeer auction house, claimed to be the largest building in the world, covering 35 hectares. Twelve million flowers and one million pot plants are sold each day! Road transport is used for local markets, but air transport from Amsterdam and Rotterdam airports is also used. Dutch flowers and pot plants are now flown all over the world. The Netherlands has over 63% of the world export market for flowers.

85

4.3 Farming In Mediterranean Spain

1 **A** shows the climate of Alicante, in southern Spain.

  a What is the maximum average monthly temperature?

  b What is the minimum average monthly temperature?

  c What is the range of temperature?

  d What is the total rainfall?

  e What is the total rainfall between May and October? What percentage of the total annual rainfall is this?

  f What problems does such a climate pose for a farmer? (Hint: think of the water needs of plants)

  g What advantages might this climate have for a farmer? (Hint: think of ripening of crops)

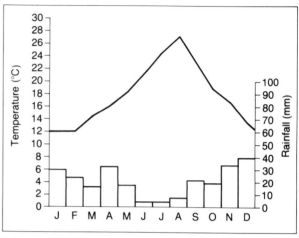

A The climate of Alicante

Long dry spells and high temperatures during the summer cause water shortages. Lack of water is a major problem for the farmer. It reduces crop **yields** (the amount produced) and limits the types of crops which can be grown. Traditional Spanish farming (known as 'dry farming' because it requires little or no irrigation) concentrates on crops such as wheat, olives and vines, which can survive the summer drought.

Most of the land in Spain is in the hands of a few very wealthy **landowners**. In the southern region of Andalusia, for example, 80% of the land is owned by just 10% of the region's landowners. Many of the people who work on the land are **tenant farmers** (they rent their farms from landowners) or landless labourers (who have no land of their own). A government land reform programme is slowly breaking up the big estates.

Some areas in southern Spain have a much more varied and prosperous agriculture. Water is the key. With a guaranteed water supply the sunshine and high temperatures allow rapid growth and ripening of plants. On Spain's narrow coastal plain, rivers provide vital water and **alluvium** (light, fertile soils.) The water is carried in irrigation channels and aqueducts to small, intensively farmed fields called **huertas**. The huertas grow fruit and vegetables, especially oranges, lemons, tomatoes and onions. In recent years several large new **irrigation** schemes have been built involving the construction of reservoirs.

2 What do the following keywords mean: crop yield; tenant farmer; landless labourer?

3 What makes wheat, olives and vines suitable for farming in southern Spain?

A major new development in Spanish agriculture is the market gardening area of El Ejido near Almeria (**B**). Since 1970 the area around El Ejido has been transformed into a landscape of plastic and polythene (**C**). Plastic greenhouses cover 12 000 hectares (for comparison, there are only 2200 ha of greenhouses in the whole of the UK). Over 15 000 smallholder farmers own greenhouses. The average size of holding is only 0.74 ha. Hundreds of thousands of tonnes of tomatoes, cucumbers, peppers, beans, cabbages, lettuce, melons and courgettes are grown in the hothouse

Skills	interpreting climate graph
Concepts	crop yields, land ownership, irrigation
Issues	land ownership

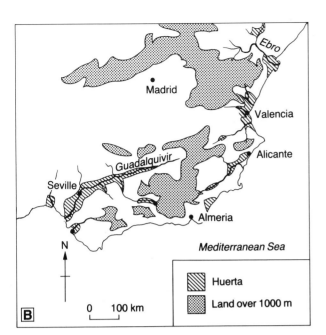

Crop	Production (thousand tonnes)	Position in World
Olives	3418	1st
Melons	751	2nd
Tangerines	868	2nd
Peaches	527	3rd
Apricots	155	3rd
Grapes	5569	3rd
Peppers	535	4th
Pears	489	4th
Tomatoes	2553	6th
Oranges	1310	6th
Onions	1005	6th
Lemons	283	7th
Cucumbers	258	8th

D Production of major fruit and vegetables in Spain and world ranking

atmosphere of the plastic greenhouses. The greenhouses are irrigated from deep wells tapping huge underground supplies of water. The crops are sent by lorry to the supermarkets of northern Europe. Over 200 000 tonnes were exported from Almeria in 1986, earning £200 million. **D** shows the production of major fruit and vegetables in Spain, and its world ranking.

The greenhouses have made El Ejido one of the richest towns in Spain. Over 12 000 migrant workers add to the town's population of 36 000. The rapid growth of the area has not been without problems, however:

● Many of the migrant workers are paid low wages and live in poor housing.

● Some of the water supplies have been contaminated with sea water because too much fresh water has been extracted.

● The plastic greenhouses and irrigation systems are very expensive to build and maintain; high winds can break the plastic and it has to be replaced every two to three years.

● There are fears that the chemicals used on the plants are trapped inside the greenhouses and make the workers ill.

4 Name the six most important fruits and vegetables (by weight) grown in Spain.
5 a Where is El Ejido?
  b How is the El Ejido area able to produce large amounts of fruit and vegetables?
  c What are the problems of the rapid growth of El Ejido?

4.4 Wealth from Industry – Antwerp

Antwerp is Belgium's leading seaport, with a population of 500 000 people. 80 million tonnes of cargo pass through Antwerp's 100 km of quays each year. Ships of up to 80 000 tonnes can sail 80 km up the River Scheldt from the open sea to Antwerp docks.

Antwerp has become one of the world's largest single industrial sites. Over 100 000 people work in the area. Many of the industries are based on the imported raw materials which come to the port:

- Sugar refining
- Flour milling
- Chocolate and biscuit manufacture
- Cigar, cigarette and tobacco production
- Vegetable oil processing – making margarine and soap
- Cutting and mounting of diamonds
- Motor vehicle assembly from imported parts (General Motors and Ford)
- Shipbuilding and ship repair
- Copper refining

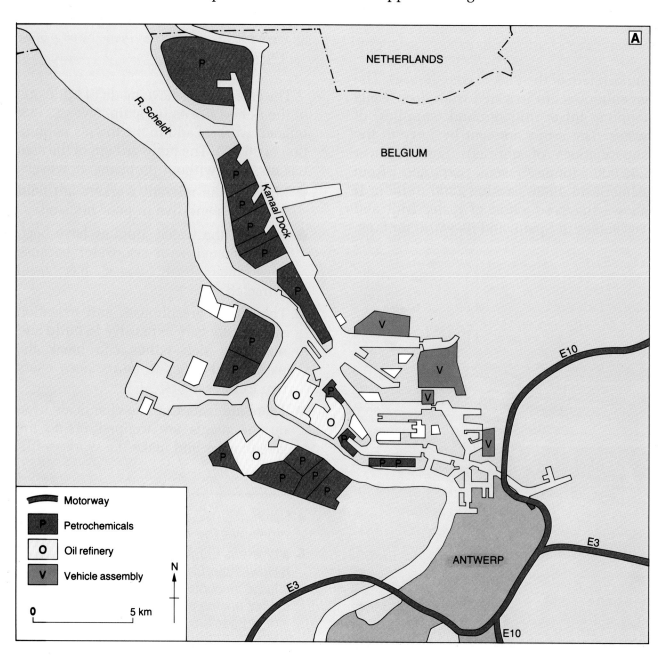

The **port industries** in the list are concentrated in and around the city of Antwerp itself (**A**). The city's most important port industry has developed downriver. Dozens of **oil refining** and **petrochemical** plants stretch along the River Scheldt and the Kanaal Dock between Antwerp and the border with the Netherlands. World famous chemical companies including Esso, BASF, Bayer, Monsanto and 3M have plants at Antwerp (**B**). They make plastics, fertilisers, synthetic rubber and textiles and many other chemical products.

The income from industry and from the port itself (**C**) means that Antwerp makes a major contribution to Belgium's wealth. Why is Antwerp such a successful location for industry? There are several factors involved:

1 The speed and efficiency of Antwerp's dockers and port authorities. Records for unloading and loading ships are regularly broken.

2 The large flat sites available along the banks of the River Scheldt.

3 The complex network of canals, roads and railways which link Antwerp with the rest of Belgium and beyond. The Scheldt-Rhine canal, opened in 1975, links Antwerp with the River Rhine and carries over 25 million tonnes of cargo per year. Antwerp is at an important motorway crossroads where the E10 Brussels – Amsterdam motorway meets the E3 Paris – Ruhr motorway.

Imports	%	Exports	%
Iron ore	28	Iron and steel	30
Coal	16	Petroleum products	26
Oil	14	Chemicals	22
Copper	5	Fertilisers	8
Steel	3	Others	14
Vegetable oils	3		
General cargo	25		

C The Trade of the Port of Antwerp

The advantages of Antwerp are so great that the oil refining and petrochemical industry has continued to grow despite the fact that large oil tankers cannot enter the Port of Antwerp. A crude oil pipeline has been built linking the oil terminals at Rotterdam with Antwerp.

1 Where is Antwerp?

2 Study **A**.
 a How far is central Antwerp from the Dutch border?
 b How are the older docks different from the more modern docks?
 c Why have the oil refineries and petrochemical industries been built downriver from central Antwerp?

3 Make lists of the port industries of Antwerp under the following headings:
 a Industries based on imported food products;
 b Industries based on imported minerals;
 c Industries based on imported components.

4 Study **C**.
 a Draw two divided bars to show Antwerp's imports and exports. Let both bars be 100 mm long.
 b What is the main difference between the types of goods imported into Antwerp and the goods exported? How do you explain this difference?

5 Design a poster advertising the advantages of Antwerp as an industrial site. Make the poster as colourful and as informative as you can. Include a sketch map showing Antwerp's situation in North West Europe.

4.5 Aerospace

Flying high across the sky is Britain's newest jet airliner, the British Aerospace 146 (**A**). It is one of the most advanced airliners in the world, able to fly from runways as short as 1000 metres. It is also one of the quietest jet aircraft, a distinct advantage for those people living near airports from which 146s fly!

Modern airliners are expensive and complicated to build. There are nearly 3000 parts, or **components**, in a BAe 146. They are produced in almost a hundred different places. **B** shows where the main components are manufactured and where **sub-assembly** of major sections takes place. The components and sub-assemblies are taken to Hatfield in Hertfordshire for **final assembly** and flight testing. **C** shows 146s on the Hatfield **assembly line**. It takes 15 weeks to produce the finished airliner. If you want to buy one you will have to find £12 million! 146s have been bought by airlines throughout the world, and the Royal Flight of the RAF carries members of the royal family in two 146s.

A

airliner, the Jetstream commuter airliner and the series of Airbus wide-bodied airliners. Military aircraft, guided missiles and satellites are also built. British Aerospace is one of the most important manufacturing companies in Britain. 75 000 people work in 23 factories located throughout the country.

As you can see from **B**, factories in Sweden and the USA are also involved in the manufacture of the BAe 146; it is a **multi-national** programme. This is usually the case in today's aerospace industry. The cost of developing new aircraft is so high that small countries such as Britain can no longer afford to build them alone. The best example of a multi-national airliner programme is Airbus Industrie. This was set up by companies from France, West Germany, Great Britain and Spain in order to build the wide-bodied Airbus 300 airliner. The A300 has been such a success that two new airliners have been built, the A310 and A320. Other countries have also become involved including Belgium, Italy and the Netherlands. By working together these European countries have been able to compete successfully with the giant US aircraft manufacturers.

---

1 What type of aircraft is the BAe 146?
2 Study **B**. Make a keyword plan to show where the main parts of the BAe 146 come from. Start with a central box:

> BAe aircraft
> Hatfield

Around it write the **name** of each part and where it comes from. Draw lines to link the parts with the central box.
3 **a** What is meant by the following terms: i) component; ii) sub-assembly; iii) assembly line?
  b Name three other products manufactured on an assembly line.

The 146 is just one of British Aerospace's products. Other civil air projects include the 125 executive jet, the ATP turbo-prop

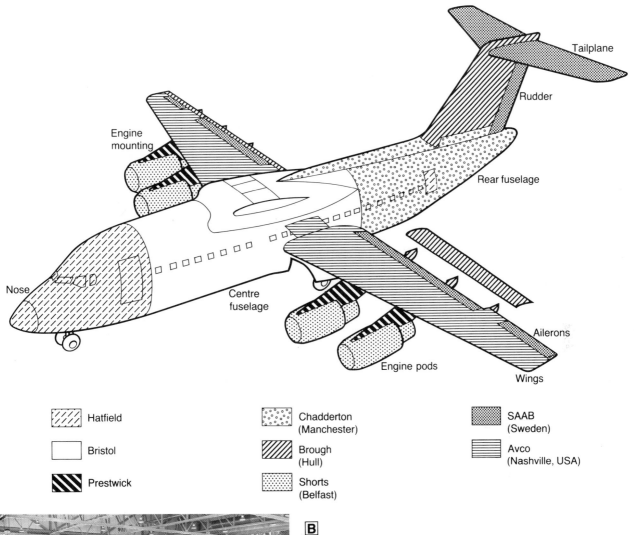

Labels on diagram: Tailplane, Rudder, Rear fuselage, Engine mounting, Nose, Centre fuselage, Engine pods, Ailerons, Wings

Key:
- Hatfield
- Bristol
- Prestwick
- Chadderton (Manchester)
- Brough (Hull)
- Shorts (Belfast)
- SAAB (Sweden)
- Avco (Nashville, USA)

B

4 What is meant by a multi-national aircraft programme?

5 Why have multi-national programmes become important in the manufacture of aircraft?

6 The table below shows how the money for the A300 Airbus project was shared between four European nations:

Nation	Share (%)
France	38
West Germany	38
Great Britain	20
Spain	4

a Construct a pie graph to illustrate these figures (remember that 1% = 3.6°).

b Name three other nations which have joined Airbus Industrie for the A310 and A320 programmes. Why did these nations join the company?

7 The USA dominates the world aerospace industry. Why do you think this is so?

91

4.6 Clothes for the world

Think about the clothes you are wearing: school uniform? Jeans? Most of us own many clothes. People usually buy clothes from shops. Some can afford clothes which are 'made to measure' especially for them.

Haute couture (high fashion) is the 'top' of the **fashion industry**. Designers and fashion houses design and make exclusive garments for the very rich and the very fashionable. Only a few of every design are made, and the clothes are very expensive. London, Paris and New York are centres of haute couture.

Wholesale manufacture is the largest sector of the clothing industry. Manufacturers copy some of the ideas of haute couture to suit the tastes of the general public. The clothes are sold through High Street shops (**retail**) such as C&A and Marks and Spencer.

A shows some of the jobs in the clothing industry.

---

1 **a** Make a list of the clothes you need for school.
  b Which lessons need special clothes? Why?
  c Make a list of jobs which require special clothes, and why.
2 What does 'haute couture' mean?
3 Where are the centres of world fashion?
4 What does **a** wholesale manufacture and **b** retail mean?

Laura Ashley

The firm of Laura Ashley started in an attic in Pimlico (London) in 1953. It is now a world-famous company. The company designs, manufactures and sells its own clothes. Most Laura Ashley clothes are made from natural fibres like cotton or wool.

In the 1970s the company **diversified** (started making other goods). It made items

Designers
Factory Manager
Warehouse Staff
Accountants
Pressers
THE CLOTHING INDUSTRY
Inspection
Sewers
Stock Controller
Pattern Cutters
Bundlers (sorting parts of the garment)

A

Skills	keyword plan, mapwork, giving talk
Concepts	wholesale/retail, world market, distribution centre
Issues	world trade/world markets

C

like curtains and sheets, and household accessories such as paint and tiles.

The largest Laura Ashley factory is in New Town (Powys, Wales) (**B**). It produces 10 million metres of fabric a year. There is a large distribution centre at Helmond in Holland. This supplies goods to shops in Europe. Another distribution centre in Mahwah (New Jersey, USA) supplies the USA and Canada. Goods are transported by road, by rail, by sea and sometimes by air.

Laura Ashley died in 1986, but the company carried on her name. There are now nearly 300 Laura Ashley shops around the world (**C**). There is also an international mail-order business.

---

5 Where and when did the firm of Laura Ashley start?

6 Design a poster to show what Laura Ashley manufactures.

---

7 Why is transport important to manufacturers like Laura Ashley?

FURTHER WORK

● Prepare a three-minute talk on the Laura Ashley company for your group. You could work in a small team to do this. Explain in your talk what the company does, how it started, why the clothes appeal to people and where they are sold. You could make and use posters or overhead transparencies to illustrate your talk. If your school has a tape recorder or video camera your talk could be recorded.

● If you were able to record your talk then play it back. Think about how you could improve your presentation. Discuss this method of presenting your work with your group and teacher. You could ask to do this again with other units in this book.

4.7 Tourism in the Mediterranean – Majorca

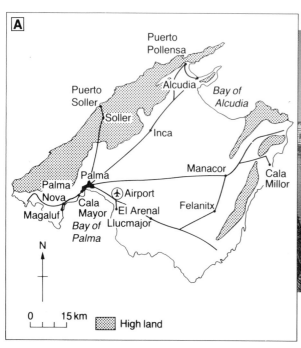

As your holiday jet descends towards Palma Airport you gain a spectacular view of the majestic mountain chain which rises sheer from the deep blue waters of the Mediterranean Sea. The mountains flash by and are replaced by a broad, fertile plain which stretches away to the far shore of the island (see **A**). As the airliner turns into its final approach the view is dominated by the hotels and apartments ringing the Bay of Palma **B**, and by the broad white beach. You are one of over four million tourists flying into Palma, Majorca, every year.

Majorca is the Mediterranean's most popular holiday island. Most of the tourists come on holidays offering a complete package with a single all-in price (a **package tour**). **C** shows the climate of Majorca. The tourists are attracted by the sunshine (an average of ten hours a day during the summer) and the heat (over 25°C on average during July and August). The mild winters encourage tourism all year round. Most

tourists stay in the resorts around the Bay of Palma: Magaluf, Palma Nova, Cala Mayor, El Arenal and the city of Palma itself.

Palma Nova has grown from open fields into one of Majorca's largest resorts within the space of a decade. A broad beach is backed by pine-covered hills. High-rise hotel and apartment blocks now obscure the view. During the day the beaches are packed with sun-seekers. In Palma Nova you can eat British fish and chips, German sauerkraut, Danish open sandwiches. There are discos, cafes, pubs, shops, clubs and cinemas (**D**). Few Spaniards visit the resort, but thousands work there.

Few who work in Palma Nova actually live there. Most prefer to live in the more traditional Spanish settlements on the island. Most have come from other areas where work is scarce and only stay for the holiday season. Tourism brings its problems – noise, litter, violence, drunkenness, theft, and the swamping of local culture and traditions by foreign ideas and customs. Many of the jobs in tourism are unskilled and poorly paid.

Skills	drawing bar graph, designing poster
Concepts	package tour, tourist development/problems
Issues	tourism

Month:	J	F	M	A	M	J	J	A	S	O	N	D
Temperature (°C):	10	10	12	15	17	21	25	24	23	18	14	12
Rainfall (mm):	40	25	30	35	20	5	0	10	20	20	30	40
Days of sunshine:	15	14	16	19	20	24	29	26	20	17	15	14

C The Climate of Palma, Majorca

Most of the tourists visiting Majorca spend their time on the beach, perhaps with an excursion for a donkey ride or a wine tasting. Those who leave the coast to travel inland discover areas less affected by tourism. There are more peaceful resorts on the north and east coasts of Majorca such as Puerto Pollensa (E). In the mountains there are remote villages which seem to belong to a different world from the noise and bustle of Palma Nova.

In 1986 46 million tourists visited Spain. Earnings from tourism amounted to £8000 million, over 5% of Spain's Gross National Product. Tourism is thus vital to Spain. Most Spaniards are prepared to put up with the problems that come with tourism, but some are worried about the effects on Spanish culture and traditions.

D

E

1 Where is Majorca?

2 a Where is Palma Nova?
  b What was Palma Nova like 20 years ago?
  c How has Palma Nova altered over the last 20 years?

3 Design a poster advertising the attractions of Majorca for a holiday. Try to make the poster appeal to as many different tastes as possible.

4 C shows the climate of Palma, Majorca.
  a Draw a bar graph to show the rainfall figures and a line graph to show the temperature figures.
  b What is the total rainfall?
  c What is the range of temperature?
  d Which would be the best month to take a holiday in Majorca?

5 What do you think are the advantages and disadvantages of tourism to Majorca?

4.8 Winter Holidays – Italian Alps

Research a ski holiday in Bardonecchia

Bardonecchia is a lively town. Its narrow main street, the Via Medail, is good for shopping. There are interesting churches. The land surrounding the town is farmed and dairying is important. A main road runs above the town linking Italy with France through the Frejus Tunnel. The Paris/Rome railway also passes through the town.

Looking over Bardonecchia towards the Campo Smith ski slope area. The slopes there are more wooded than the Jafferau slopes.

You are planning to go skiing for one week in the Italian Alps, starting on March 1st. Before you book you have to **research** the holiday and **cost it**.

---

1 Use the information in this unit to complete the following keyword plan about Bardonecchia.

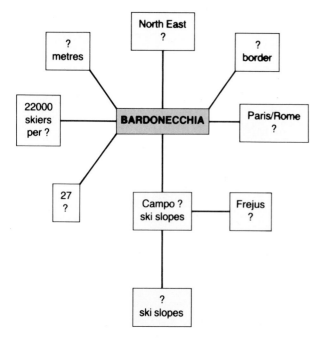

2 Apart from skiing, what else does the town offer?

The railway station at Bardonecchia with the town and mountains behind.

---

3 Use the **Account sheet** to calculate how much your holiday would cost in **a** the apartment **b** the Riky Hotel. Which one will you choose? State your reasons.

Now add on the other costs: **c** insurance; **d** travel from Luton or Manchester if you wish to use one of these airports; **e** one week ski hire; **f** ski school; **g** a lift pass; **h** anything else you think you will need for the week.

ACCOUNT SHEET
March 1 (7 days)

Accommodation	Basic price	Insurance	+ from Luton	+ from Manchester	1 week ski hire	Ski school	Lift pass	Extras	Total cost
APARTMENT	£168								
RIKY HOTEL					£8.00				

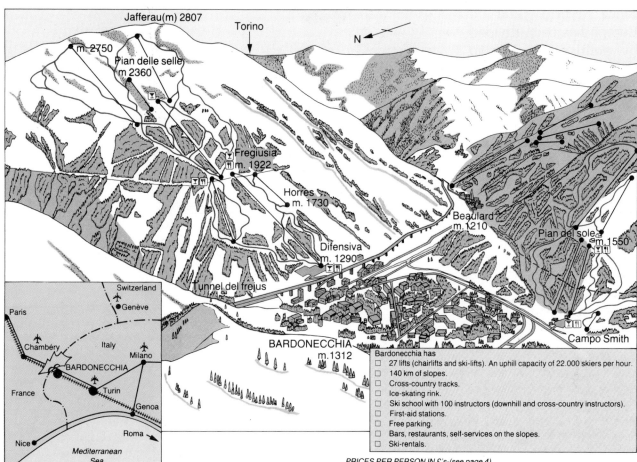

4 What would you look forward to most about your holiday?

5 You work for a ski holiday company. You have three minutes to 'advertise' Bardonecchia. Prepare your presentation. You may use a script, a tape-recording or a video.

FURTHER WORK

● Draw a *fieldsketch* (line sketch) of the village and ski slope area (or use the outline in the *Activity Pack*). Or, mark in and label some of the ski lifts.

Bardonecchia has

☐ 27 lifts (chairlifts and ski-lifts). An uphill capacity of 22.000 skiers per hour.
☐ 140 km of slopes.
☐ Cross-country tracks.
☐ Ice-skating rink.
☐ Ski school with 100 instructors (downhill and cross-country instructors).
☐ First-aid stations.
☐ Free parking.
☐ Bars, restaurants, self-services on the slopes.
☐ Ski-rentals.

PRICES PER PERSON IN £'s-(see page 4).
REMEMBER TO ADD INSURANCE 4–9 NIGHTS
£17.95; 10–14 NIGHTS £20.95

Hotel & Board Arrangements	Bardonecchia Apartments (SC)		Riky (HB)	
Hotel Code	WBX		WBR	
Nights in Hotel (p. 169)	7	14	7	14
Party size (excl. infants)	8		–	
Dec 21	178	221	298	453
Dec 28	190	201	310	433
Jan 4	135	161	255	393
Jan 11, 18	131	161	251	393
Jan 25	131	174	251	406
Feb 1	141	182	261	414
Feb 8	153	192	273	424
Feb 15	161	202	281	434
Feb 22	188	217	308	449
Mar 1	168	201	288	433
Mar 8, 15, 22	142	174	262	406
Mar 29, Apr 5	138	168	258	400

SKI PACKS
Book in UK

Code	Ski Pack	Price
A	1 week Ski Hire	£8.00
B	1 week Ski School (18 hrs)	£28.00
C	1 week Lift Pass	£38.00
L	1 week Learn to Ski	£79.95
W	1 week Boot Hire	£3.40
D	2 weeks Ski Hire (36 hrs)	£12.00
E	2 weeks Ski School	£56.00
F	2 weeks Lift Pass	£76.00
X	2 weeks Boot Hire	£8.00

● Ski Pack L - see page 8 for details

SUNDAY FLIGHTS TO TURIN.
Departure codes and supplements per person

GATWICK	3282 £0	LUTON	3283 £3	MANCHESTER	3284 £17
SKI DRIVE/ACCOMMODATION ONLY				3328 **Deduct £75**	

4.9 The Great Divide – Eastern Europe

Since World War II Europe has been divided by politics (**A**). Eastern Europe has been strongly influenced by the Soviet Union. Soviet troops are stationed in Eastern Europe. Western Europe has been strongly influenced by the USA. American troops are stationed in Western European countries.

Much of Eastern Europe's trade is with the USSR. People in the East do not have such a high standard of living as those living in Western Europe. It is not so easy to travel. There is little political freedom; people cannot usually vote for candidates opposing the government. The governments are **communist**.

Communist countries believe in state ownership of factories, farms, transport and education. Governments set targets for production and they control the economy. In the countries of Western Europe and North America there is mainly **private** ownership of land and industry. This system is called **capitalism**. The state does not play such an important role in the economies of capitalist countries.

More recently some private marketing has been allowed in Eastern Europe. Today Hungary and Romania have the most private trading of all the communist countries. This kind of economy is called **market socialism**.

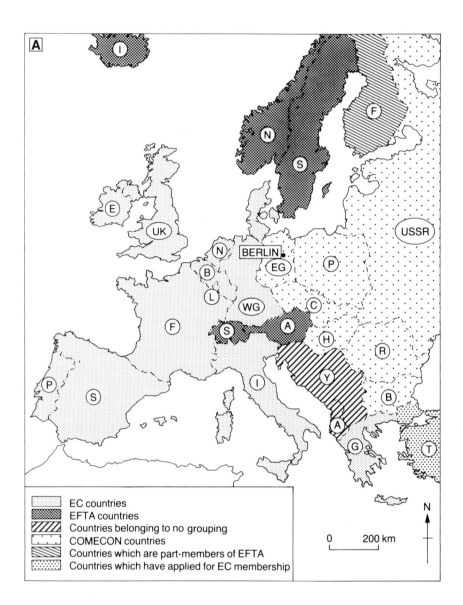

A

EC countries
EFTA countries
Countries belonging to no grouping
COMECON countries
Countries which are part-members of EFTA
Countries which have applied for EC membership

0   200 km

N

1 a Which country influences Eastern Europe?
 b Which influences Western Europe?
2 a What do communist countries believe in?
 b What do Western countries believe in?
3 Study **A** and the details about the groupings of countries in Eastern Europe.
 a What do the initials COMECON stand for?
 b What is the Warsaw Pact?
4 Using your atlas name the countries in each of the groupings shown in **A**.

Skills	atlas work
Concepts	country groupings: communist/capitalist, East/West, political divisions
Issues	division, barriers

5 Draw out keyword plans for the West and the East. Include the political and economic terms eg (West – Europe, North America, private ownership, limited state role; East – Europe, USSR, state ownership, state control, market socialism)

Hungary

If you visited Hungary you would not be very aware of the 'Great Divide'. This **landlocked** country (with no coastline) has 11 million people. Shops in Hungary are full of goods – not always the case in other Eastern European countries and the USSR. Hungary has spent money expanding its industries. It now exports power station boilers; transformers; railway locomotives; buses; and river barges.

Hungarian industries have joined up with Western European companies, including West German, Austrian, Swiss and British firms. Hungarian and West German **partnerships** have recently sold buses to Venezuela and power stations to Turkey and Greece.

6 Name two of Hungary's transport exports.
7 How has Hungary changed from a purely socialist state?

Berlin

The city of Berlin is situated inside East Germany. If you visited Berlin you would definitely be aware of the Great Divide. After World War II the city was divided up among the French, British, Americans and the Soviets (**B**).

One day in August 1961 East German troops in the city rolled out coils of barbed wire along the east/west boundary. The next day a wall of concrete blocks was laid. Today the Berlin Wall is a double one, enclosing a 'death strip' with dog runs, trip wires, searchlights, booby traps and armed guards (**C**). The Berlin Wall was built to stop East Germans escaping to the West. It is a permanent reminder that there is a Great Divide between East and West Europe. Why does it remain? Is it really necessary?

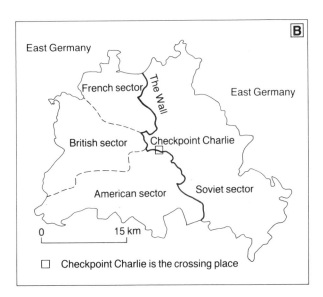

Checkpoint Charlie is the crossing place

8 Why was the Berlin Wall built?
9 Why is it evidence that there is a Great Divide between East and West Europe?

FURTHER WORK

● Make a list of events happening in Eastern Europe that are in the news at the moment.

4.10 Further Afield – The USA and the USSR

What do you think of when the United States of America (USA) and the Union of Soviet Socialist Republics (USSR) are mentioned together? You may think of possible nuclear war, space flight, spying, arguments or mistrust.

The USA and the USSR are **Superpowers**. The United States controls nuclear weapons in the West and the Soviet Union controls them in the East (**A**).

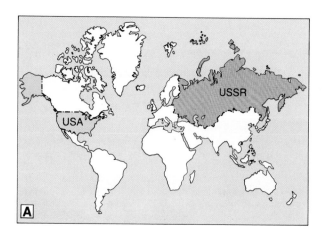

These two powerful nations have similar geography. They both have remote mountain areas, desert lands, cold empty lands rich in minerals, fertile farming regions and major centres of population and industry. Maps **B** and **C** show other facts about the two countries.

Why are they different?

Throughout history people have grouped themselves together. These groups of people (countries) have often agreed to defend each other if attacked. In this century two ways of running a country have grown and spread worldwide. They are **Capitalism** and **Communism** (see Unit 4.9).

The United States can be said to be the Superpower defending the capitalist world; the Soviet Union defends the communist world. This great political difference underlies all the conflicts between them.

Capitalists believe in the freedom of people to decide what they want.

Communists believe that there should be no private ownership and no social classes.

The USA sees Communism as a threat to Capitalism, and so will often support other countries to prevent them becoming communist. The USSR will support any countries trying to become communist.

The world is very concerned that the two Superpowers might try to settle their differences by going to war. Both have large supplies of nuclear weapons and could accidentally or deliberately use them. Some people believe that if this happens there will be no winner but the whole world may suffer as a result. At present both sides keep nuclear weapons as a **deterrent** to put off the other side from attacking. The Superpowers and their allies (supporters) keep the **balance of world peace**. Most of the world hopes they will keep talking to gain greater understanding of each other and hopefully live in peace.

---

1 What do the letters USA and USSR stand for?

2 In what ways are the two countries similar?

3 Why are they called 'Superpowers'?

4 Use maps **B** and **C** to explain why the two nations are so powerful.

5 What is **a** Capitalism? **b** Communism?

6 Why is the world worried about the two countries and their differences?

7 What is **a** NATO, **b** The Warsaw Pact? Which countries belong to each organisation (see **B** and **C**)?

Skills	survey work, discussion
Concepts	East/West, superpowers, capitalism/communism
Issues	world peace

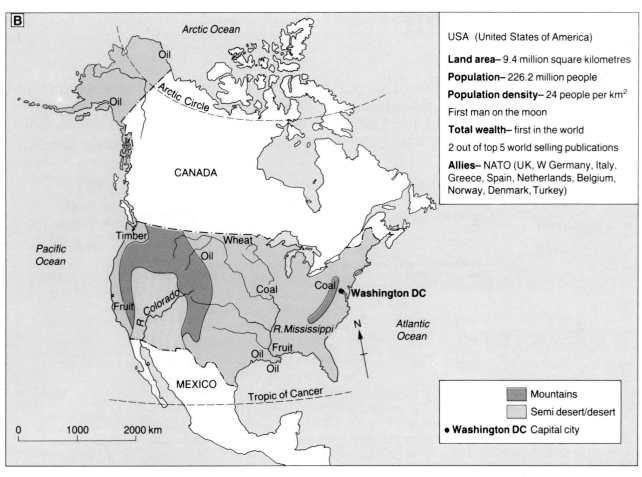

B

Arctic Ocean

Oil

Arctic Circle

Oil

Oil

CANADA

Pacific
Ocean

Timber

Wheat

Oil

Coal

Coal

Washington DC

Fruit

R. Colorado

R. Mississippi

N

Atlantic
Ocean

Oil

Fruit

Oil

MEXICO

Tropic of Cancer

0   1000   2000 km

USA (United States of America)

Land area– 9.4 million square kilometres

Population– 226.2 million people

Population density– 24 people per km^2

First man on the moon

Total wealth– first in the world

2 out of top 5 world selling publications

Allies– NATO (UK, W Germany, Italy, Greece, Spain, Netherlands, Belgium, Norway, Denmark, Turkey)

	Mountains
	Semi desert/desert
● **Washington DC**	Capital city

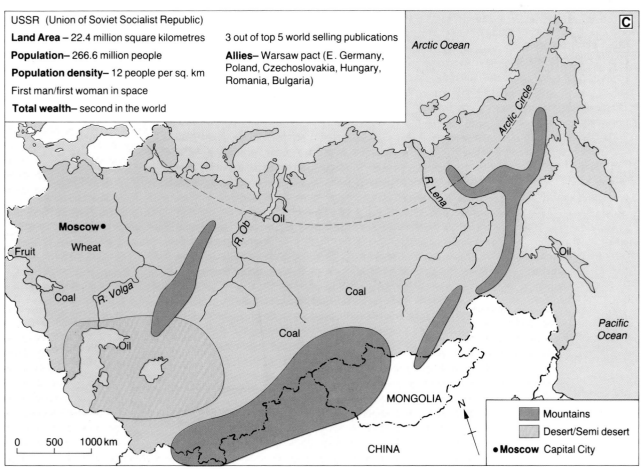

C

USSR (Union of Soviet Socialist Republic)

Land Area – 22.4 million square kilometres

Population– 266.6 million people

Population density– 12 people per sq. km

First man/first woman in space

Total wealth– second in the world

3 out of top 5 world selling publications

Allies– Warsaw pact (E. Germany, Poland, Czechoslovakia, Hungary, Romania, Bulgaria)

Arctic Ocean

Arctic Circle

R. Lena

Moscow ●

Fruit

Wheat

R. Ob

Oil

Oil

Coal

R. Volga

Coal

Coal

Oil

Pacific
Ocean

Oil

MONGOLIA

N

	Mountains
	Desert/Semi desert
● **Moscow**	Capital City

CHINA

0   500   1000 km

101

4.11 Rust Belt

Have you heard of the Great Lakes of North America? They are one of the famous features of the continent. Find this huge waterway system in an Atlas. The Great Lakes are an important feature of the Mid-West Industrial Region of the USA, one of the most important industrial regions in the world. Why did industry develop here?

A

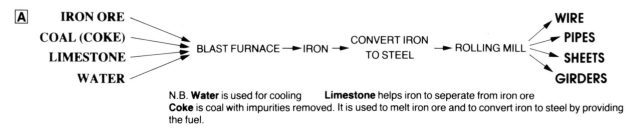

N.B. **Water** is used for cooling   **Limestone** helps iron to seperate from iron ore
Coke is coal with impurities removed. It is used to melt iron ore and to convert iron to steel by providing the fuel.

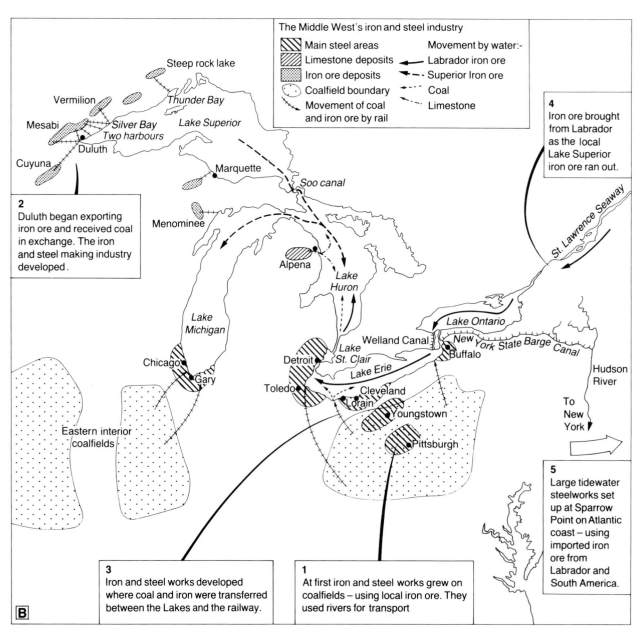

B

Skills	mapwork, groupwork
Concepts	iron and steel production, industrial decline, rust belt
Issues	unemployment

A shows what is needed to make iron and steel. All of the raw materials are found in the Mid-West so it became a centre of the iron and steel industry (see **B**). This attracted manufacturers of vehicles, ships, locomotives and many other products to the area.

The fall

Over the last 20 years there has been a steep decline in traditional manufacturing regions of Britain (see Unit 4.6 of Book 1). This is also true in the Mid-West Industrial Region. Reasons for the decline include

● many of the industries which used steel are themselves in decline because their products are no longer needed (for example shipbuilding);

● countries who used to import US steel now make their own;

● foreign made steel and goods made from steel are cheaper than those produced in the USA (see Unit 4.15);

● new materials such as plastics have replaced steel;

● local raw materials are running out, so steelmaking has become more expensive and uneconomic.

Cleveland, on the shores of Lake Erie in the state of Ohio, is typical of what has happened to the old manufacturing centres of the USA. **C** gives some of the reasons for the decline of Cleveland.

Today, what was once a great manufacturing region is becoming a **rust belt**. Job losses are high in iron and steel and the industries it supported. Streamlining production by introducing robots has cut jobs. Unemployment is serious. Every year the number of people needing help from the government doubles. Cleveland has an unfortunate image of being a city which is slowly dying. In the past ten years 60% of the people have left. Derelict land and buildings increase as people move elsewhere in search of work. Unit 4.12 looks at where they go.

High wages, taxes put off industrialists

Less demand for goods made from steel

As resources run out it is expensive to import raw materials. The cost of steel rises, making it too expensive to buy

Newer materials (plastics) are replacing steel

REASONS FOR THE DECLINE OF CLEVELAND

Coal mines closing in northeast USA - new ones are 2000 km west

The south and south west of USA attract industry because they can make higher profits

Manufacturers using steel have moved or closed down

C

1 Complete the map in the *Activity pack* using map **B**.

2 Copy flow chart **A** and use map **B** to add where the raw materials come from.

3 What manufacturing industries used iron and steel?

4 The Great Lakes is an important waterway system. It is 1200 kilometres from Lake Ontario in the east to Lake Superior in the west. Why did Cleveland benefit from being located on the Great Lakes?

5 Why have iron and steel centres like Cleveland (see **C**) declined?

6 Why do you think that people refer to the Mid-West Region as 'The Rust Belt'?

FURTHER WORK

● Imagine you are a Cleveland steelworker. After 15 years you lose your job. Discuss with your group how each of you would feel if you were the person involved. How would it affect your family?

● Why do you think some people choose to stay in Cleveland when they lose their job whereas others leave the city?

● How would the movement of people away from Cleveland affect: local shops? local schools? community spirit?

PEOPLE'S LIVES AND WORK

4.12 Sunbelt USA

In Unit 4.11 you read about the decline of the old traditional manufacturing region of the Mid-West USA. This is a common problem in the developed world. New industries tend to be **footloose**. They are free to locate where they want to, as they are not restricted by the need for lots of heavy raw materials. The finished product is light and easily transported by road. Phoenix in Arizona is typical of the fast-growing cities which attract the new footloose industries. What makes Phoenix part of the second fastest growing industrial area in the USA?

The Sunbelt

In recent times there has been a north-south movement of people in the USA. The southern states have become very attractive to industrialists and investors. These people want to make high profits and the **sunbelt states (A)** can help them.

Why do these areas help industries make high profits? The reasons include:
● Cheap land
● Lower wages for workers
● Companies pay lower taxes
● Trade unions are restricted
● Attractive climate/scenery
● Good transport links

Phoenix, Arizona

The Phoenix metropolitan area has a very hot and sunny climate where the average temperature is 29°C and annual rainfall is only 176 millimetres. It is located in the semi-desert countryside of Southern Arizona.

The city of Phoenix is spreading outwards and at present measures 80 kilometres across (see **B**). Most of the building is low level, apart from the tall skyscrapers of the city centre. The city is well linked to the US Highway system and has good rail and air links. The city provides financial assistance

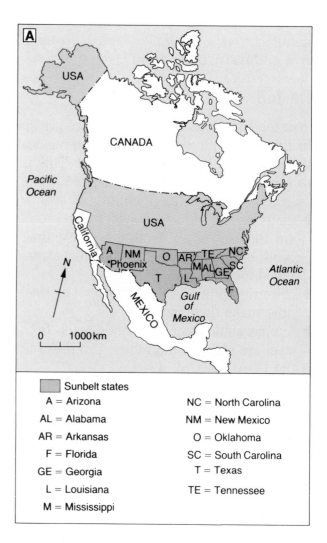

Sunbelt states
A = Arizona
AL = Alabama
AR = Arkansas
F = Florida
GE = Georgia
L = Louisiana
M = Mississippi

NC = North Carolina
NM = New Mexico
O = Oklahoma
SC = South Carolina
T = Texas
TE = Tennessee

for new industry and at present has over 2500 firms. Manufacturing is the major income producer and the leading manufacturing industry is electronics (for example computers).

The Environment and People

Most people in Phoenix have a good standard of living and the climate helps them to enjoy a variety of leisure activities including golf, tennis and sailing.

For many newcomers from the 'Rust Belt' of the Mid-West the heat is a problem. Air conditioning is needed in cars/buildings and gardens/golf courses need water frequently. The rapid growth of the city has made heavy demands on water and electricity.

Skills	mapwork interpretation, keyword plan, empathy, problem solving
Concepts	footloose industry, sunbelt, adaptation
Issues	expansion

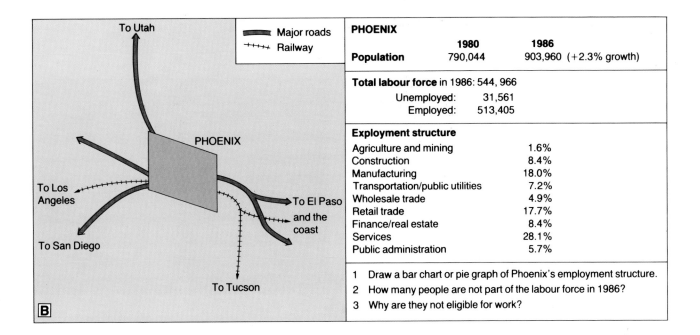

B

To Utah
To Los Angeles
To San Diego
To El Paso and the coast
To Tucson
PHOENIX

Major roads
Railway

PHOENIX

	1980	1986
Population	790,044	903,960 (+2.3% growth)

Total labour force in 1986: 544,966
Unemployed: 31,561
Employed: 513,405

Exployment structure

Agriculture and mining	1.6%
Construction	8.4%
Manufacturing	18.0%
Transportation/public utilities	7.2%
Wholesale trade	4.9%
Retail trade	17.7%
Finance/real estate	8.4%
Services	28.1%
Public administration	5.7%

1 Draw a bar chart or pie graph of Phoenix's employment structure.
2 How many people are not part of the labour force in 1986?
3 Why are they not eligible for work?

A canal has had to be built to bring water 500 kilometres from the Colorado River at a cost of $3 000 000 000. As the city expands outwards then people have further to travel to work and it becomes more expensive to provide them with the services they need, such as water. Some newcomers have found that they do not have the skills needed by the new industries – life for them is very difficult.

1 How many 'Sunbelt States' are shown in **A**?
2 Why do you think they are called 'Sunbelt' states?
3 Copy and complete this keyword plan of the reasons why Phoenix is attractive to industrialists:

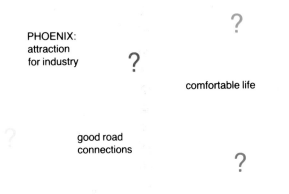

PHOENIX: attraction for industry

?
?
comfortable life
?
good road connections
?

4 Design a poster for the Arizona Department of Commerce aiming to attract industries to move from the 'Rustbelt' of the North to the 'Sunbelt' of the south. Think of a good slogan to sell the idea.

5 Imagine you are a newcomer from the north arriving in Phoenix to start work in a computer factory. Describe your feelings about your new home and the environment to someone in your group. Listen to what their ideas are. What problems did you both imagine you would face? What would your feelings and problems be if you lacked the skills needed for the new industries?

6 What problems does Phoenix face as a rapidly expanding city?

FURTHER WORK

● Imagine your group are the planners of Phoenix. You need to slow the outward growth of the city before it becomes impossible to service the outer suburbs with water, electricity and roads. Hold a meeting to discuss what you might be able to do about the problem. Present your ideas to the rest of your class. What are their views on your proposals?

4.13 Moscow and Beyond

Geographers are interested in governments and the **plans** they have for change in their country. Government **decisions** can affect everyone in the country. In some countries the government controls all decisions centrally. This means that the government thinks it knows what is best for its people and how to provide what the people need. One example is in Communist countries such as the USSR.

Moscow

The USSR is the largest country in the world. Part of it is in Europe and part in Asia. It is a **superpower** (see Unit 4.10). Moscow is the capital of this communist country (see **A** and **B**). A group of pupils wanted to find out more about the city. They went to the library and wrote to the Russian embassy in London for information. **C** is a keyword plan of the main points they discovered.

Beyond Moscow

Moscow is the economic centre of the European (western) part of the USSR. The greatest economic development has been in the western USSR. The government is concerned because a lot of valuable resources are in the east and so it is planning economic development beyond Moscow and the Central Region it serves.

In the USSR the government makes all the decisions about industries. It has decided to distribute industry widely throughout the country. Why does the Soviet government want to do this? **D** suggests some of the main reasons.

C

Capital city
8 million people
Centre for communications (road, rail, air)
Most important USSR tourist centre
TV, radio and newspaper headquarters
Centre for education (universities/libraries)
Centre for culture/entertainment (eg Bolshoi Ballet)

MOSCOW

Most people live in small flats
Largest city in USSR
Main government offices (Kremlin)
1980 Olympic games held there
Important industrial centre
One of the largest cities in the world (900 km^2)

Skills   map/atlas work, measuring, making a guide
Concepts government control, decentralisation, movement of industry, industrial development
Issues   government control

Reasons for developing industry in Siberia, Soviet Central Asia and the Far East.

1 Many resources needed by industry are found in Soviet Central Asia, for example coal, iron ore, oil and gas.

2 There are deposits of rare minerals in the east, for example gold, tin and copper.

3 Rivers can produce hydro-electric power for energy.

4 It is further away from Western Europe and possible attack from the West.

5 They are worried that China may try and take resources in the east if they are not developed.

6 Japan in the east may trade with the USSR for the resources it needs to keep its successful industries going.

7 The government wants to develop its land area and increase the general wealth and productivity of the country.

D

To achieve these aims the government has created regions called **territorial production complexes** (TPC) which are growth points. These encourage development of resources and industry in carefully chosen locations. **E** shows the distribution of TPCs.

1 Complete the map in the *Acitivy Pack* using map **E** and an atlas to help you.
2 Use the keyword plan and photographs of Moscow to make a small guide to explain to tourists why Moscow is such an important city.
3 What is a TPC?
4 How many TPCs are marked on the map?
5 How long is the Trans-Siberian Railway from Moscow to Vladivostock? (see map **E**).
6 Why do you think this railway is important to the development of the east?
7 Why might Vladivostock be important if the USSR decided to export resources to Japan?

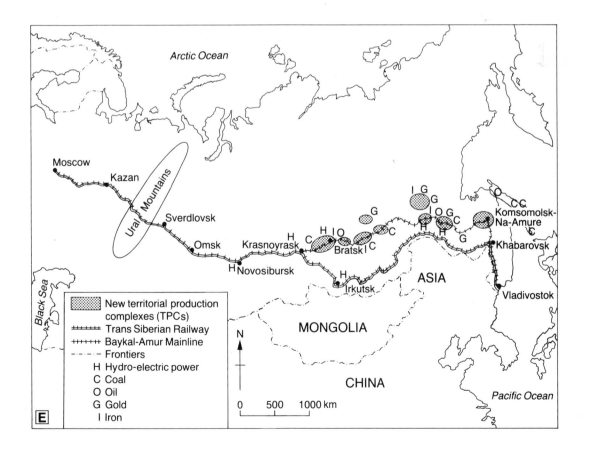

E

New territorial production complexes (TPCs)
┼┼┼┼┼ Trans Siberian Railway
++++++ Baykal-Amur Mainline
- - - - Frontiers
H Hydro-electric power
C Coal
O Oil
G Gold
I Iron

4.14 Home and Away – Culture Clash?

Culture is the way of life passed down from one generation to another. It includes the way people think and what they believe in. What happens when a traditional lifestyle begins to be replaced by a new one? This may cause **culture clash** where different attitudes meet. The traditional culture may feel threatened by the foreign attitudes as people change their way of life, forget family traditions, change their eating habits and prefer foreign television to listening to story-tellers and poetry.

Japan is becoming **westernised**, especially in its cities. People tend to wear clothes like people in Europe and America. Advertisements on television are often foreign but dubbed in Japanese. Western-style food and music is popular, particularly amongst the young (**A**).

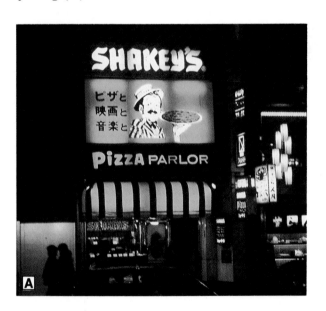

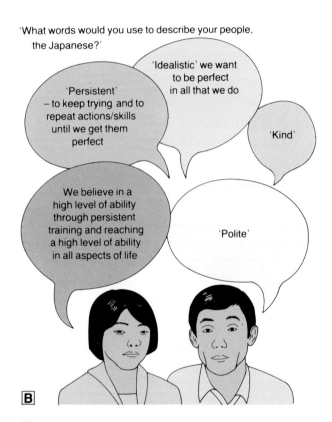

'What words would you use to describe your people, the Japanese?'

'Persistent' – to keep trying and to repeat actions/skills until we get them perfect

'Idealistic' we want to be perfect in all that we do

'Kind'

We believe in a high level of ability through persistent training and reaching a high level of ability in all aspects of life

'Polite'

B

Home

Have the Japanese lost their culture? **B** shows how the Japanese described themselves in a 'National Character Survey' in 1968. Japan may look modern, especially in the cities which are like those in North America and Europe, but the old ways still continue. The **IE** or extended Japanese family (parents, children, grandparents, relatives) still exists. The mother and children are at the centre of family life (**C**) and the eldest son is expected to follow in the father's footsteps. Young families tend to consist only of parents and children and some husbands now tend to put their families before work. Couples now walk hand in hand instead of the women following several paces behind.

Religion is still strongly part of Japanese life. 84 million people follow Buddhism and 92 million Shinto (many Japanese

---

1 What is **culture**?

2 Look at **A**. Why is this evidence of 'culture clash'?

3 If you visited Japan where would you notice that people have become 'westernised'?

Skills	*photograph interpretation, empathy, cartoon,*
Concepts	*culture, culture clash, westernisation*
Issues	*culture clash, Japanese industry in Britain*

follow more than one religion). The idea of good service and respect for the customer continues although now it is seen in fast-food places and supermarkets. So, although foreign ideas have entered Japan they tend to be adapted rather than taking over. Underneath the modern face of Japan the traditional culture is followed.

4 Copy and complete the keyword plan below.

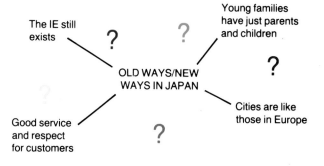

The IE still exists ?

Young families have just parents and children ?

?

OLD WAYS/NEW WAYS IN JAPAN

Cities are like those in Europe ?

Good service and respect for customers ?

Away

One of the major changes in the last 20 years has been the move by Japanese industry to locate in other countries. This has led to the Japanese attitude to work being exported. One example is the Komatsu excavator and bulldozer factory located near Newcastle in England (**D**). Workers are chosen for their loyalty and their ability to work in a team, rather than their experience.

E is part of a newspaper report on what it is like to work for Japanese-owned companies in Britain.

Mr Jones was asked about working for Japanese bosses in the factory and how it is different from British ways of working.

'The traditional attitude of British business "I'm the boss, you work for me" has been changed to "You work with us, we'll work as a team". There is equal status whether you are the manager or a fitter like me. All wear the company uniform and follow the company aim "To deliver to the customer the world's top class product". Workers are shown the finished product to understand what they are contributing to and are encouraged to suggest how to improve their team's part of the manufacture and assembly.'

E

Problems

In Japan a worker can expect to be employed for life by the one company. The company comes first. In Britain workers tend to put their family first, ahead of their job. The guarantee of a job for life cannot be made. Unlike the Japanese British workers are reluctant to work overtime for nothing. Some British workers cannot get used to the equal status of workers and bosses.

5 Design a poster for a Japanese company in Britain. The aim of the poster is to remind British workers of the company aim TO DELIVER THE CUSTOMER THE WORLD'S TOP CLASS PRODUCT. This could be a car or bulldozer.

4.15 Success Story

After World War II Japan lay in ruins (**A**). As a defeated and now occupied country it had to rebuild industry, housing, transport and the people's confidence. How could it begin? At first the United States gave money and guided the rebuilding. Small owner-farmers were encouraged, trade unions strengthened, and the power of the big industrial families was broken. The Japanese decided to concentrate on developing heavy and chemical-based industry where profits would be higher.

Expansion

Problems of lack of raw materials and lowland to build on (most of the country is mountainous) had to be tackled. As most raw materials had to be imported, factories were built on the coast. There was big investment in iron and steel, chemicals, textiles, cars and electrical goods. Later, the electronics industry was developed. The education system provided skilled workers and many were transferred from agriculture into the rapidly-expanding industries. A stable supply of imported raw materials and energy was set up. The Government concentrated on spotting future markets for Japanese goods overseas and advised industry to plan for these.

1 What was Japan like at the end of the Second World War?

2 What did the Government decide to do?

3 Most factories were built on the coast. Apart from importing raw materials, what other reasons might there be for this location? (**Clue:** Where did manufactured goods go?)

Industrial Location

Japan's major industries are located in three areas (see **B**). This string of locations is known as the 'Pacific Belt'. During Japan's process of recovery and expansion iron and steel, oil refining, chemical works and power stations were located on land **reclaimed** from the sea.

The first reclamation sites were chosen for their ease of access to the home market with good road and rail links. Later, there was inland industrial expansion around

A

B

Skills	mapwork, research
Concepts	industrial development, import/export, asembly line, high tech industry
Issues	shortage of land, pollution, Japanese imports

the edges of large cities. This consisted mainly of **assembly-line** industries, for example radio/television assembly and motor vehicles.

Recently, industry has developed in other regions, for example North Kanto and Tohoku. New industrial cities include Ouita. Special development areas have been developed, Eastern Tomakomai being especially successful.

The Key to Success

The ability to organise and the positive attitude of the workforce and management have both been vital to Japan's industrial success. The attitude of management is 'What is good for the workers is good for the company'. Employees feel loyalty to the company and pride in their work. Companies provide medical care, loans for housing and subsidised holidays. Workers discuss their jobs and how to be more efficient. Teamwork is the key to the success of Japan's industry. Management keeps workers informed of what is going on and works alongside them.

The Japanese Government is promoting knowledge-intensive high technology industries – for example advanced electronics (**C**). These industries use less energy (50% of Japan's imports by value is oil) and make big profits. Japan currently has 40% of the world market in microchips. Some industries have become depressed – for example shipbuilding, coalmining, and textiles. Pollution is a major problem and there is still a land shortage and reliance on imports of raw materials. Several countries are worried about the success of Japan's exports. The United States has mounted a campaign to persuade Americans to buy American-made goods rather than Japanese goods.

Measure of success

Today Japan is one of the leading industrial nations in the world. The country is able to adapt to changing demands in the world market. It has strengthened trading links by tending to import raw materials from countries bordering the Pacific Ocean.

---

4 Design a poster to show Japan's successes and problems. Include
   a Evidence of why Japan is successful
   b Industries in trouble and pollution
   c Attitudes/worries of other countries.
5 Complete the map in the *Activity pack*.

FURTHER WORK

● Make a list of manufactured goods in your home or school and where they were made. Compare your findings with other members of your group and find out which items were made in Japan.
● People worry about Japanese imports causing local firms to shut down. Find out why people buy Japanese goods. You could visit local shops and garages to find out why they stock Japanese goods and why customers buy them.

4.16 Megalopolis – Southern Honshu

A shows the population density of the four main islands that form Japan. Where are the most heavily populated areas? Japan has been called the 'crowded islands'. Most of its 120 million people live in a small part of the land. Find a physical map of Japan in an atlas; you will see that most of the country is highland. In fact 72% is mountains, hills or volcanoes. This means only about a quarter of the land is suitable for building. Most of this land is on the coast.

---

1 Look at **A**. What is the population density of Keihin, Chukyo and Keihanshin? (Use the key to find the value of the shading.)
2 Why do people call Japan the 'Crowded Islands'?
3 Explain why most people live on just over a quarter of the land.

The people of Japan live by making the most of their limited resources (land and raw materials available), importing raw materials and exporting manufactured goods. To keep making progress industry expands and more offices, shops and homes are built. New land has to be reclaimed from the sea for this building.

76% of Japanese people live in cities (**B**). Some cities have joined together to form **conurbations**. Three major regions of population are located on the Pacific coast of Honshu. The three – Keihin, Chukyo, Keihanshin and the areas joining them – are called the **Tokaido Megalopolis**

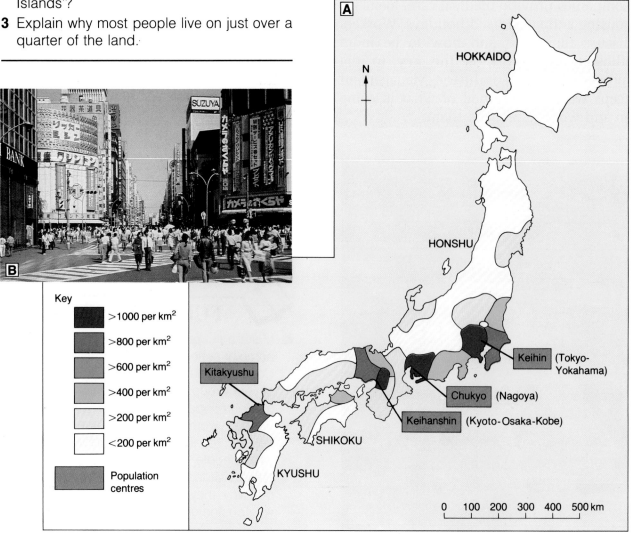

Key

>1000 per km²
>800 per km²
>600 per km²
>400 per km²
>200 per km²
<200 per km²

Population centres

HOKKAIDO

N

HONSHU

Kitakyushu

SHIKOKU

KYUSHU

Keihin (Tokyo-Yokahama)

Chukyo (Nagoya)

Keihanshin (Kyoto-Osaka-Kobe)

0 100 200 300 400 500 km

Skills	map interpretation, keyword plan, measurement
Concepts	population density, conurbation, megalopolis, commuters
Issues	problem of urban growth

(Megalopolis means a large population formed by conurbations joining together). Tokaido is the old Japanese name for a road that links cities together. With Kitakyushu on the island of Kyushu the Megalopolis has nearly 60% of Japan's people. Tokaido contains 48% of manufacturing workers.

---

4 What is the meaning of **a** conurbation and **b** megalopolis?

5 Where is the Tokaido and what is it?

Tokyo is the nerve centre of Tokaido. It is one of the world's largest cities with a population of 12 million. **Commuters** from a 50 kilometre zone around the city arrive daily for work. 27 000 000 people live in this zone where travel can take up to two hours from home to the city. Passenger traffic by road has grown since 1970 (**C**). 'Bullet' trains, among the world's fastest trains, link Tokyo to Osaka and Hokat (Kyushu). The city of Tokyo has the fourth largest subway system (underground railway) in the world. A steady improvement of the **network** of rapid transport, for example expressways (fast roads with many lanes), super-express railways and air services, has encouraged people to commute to work.

Problems

With a high concentration of people and industry, the Megalopolis suffers from chemical and noise pollution. High land prices mean that most people have small homes or flats, most without gardens; houses have an average floor area of 46 square metres. Measure the floor area of your house. Is this bigger than 46 square metres? Ueno Park (Tokyo) is one of many parks crowded at weekends with people seeking open spaces. People face long journeys to work on crowded buses and trains. Water and power supplies are under great strain to supply the population.

Rural areas are losing young people who are attracted to the city.

---

6 Copy and complete the keyword plan for Tokyo.

nerve centre of Tokaido

TOKYO

27 million people live in commuter zone

C

FURTHER WORK

● Find the area of your classroom in square metres. How does this compare with the average floor space in a Japanese home?

● How do you think the Japanese have managed to fit in transport and buildings in such limited spaces available in the cities? (**Hint:** Look at **B** and **C**).

Matrix of Concepts and Issues

UNIT NO.	1.1	1.2	1.3	1.4	1.5	1.6	1.7	1.8	1.9	1.10	1.11	1.12	1.13	2.1	2.2	2.3	2.4	2.5	2.6	2.7	2.8	2.9	2.10
Issues																							
rural life						○	⊕																
planning					○		○			○		⊕										●	
employment					○							⊕										●	
preservation/conservation					○					○													⊕
public services				○	○		○	⊕	●			●											
perceptions/viewpoints			○		⊕					●												→	→
human interference																		○				⊕	⊕
conflict					○					○	○												
pollution																							○
surplus/deficit																						○	
industrialisation		○			○				○	○	⊕	⊕	●										→
technology					○						○	○										○	○
political divisions	○	○	○	○	○						○		○									○	
prejudice		⊕								●	●												
inequality										○	⊕	⊕											
managing the environment																			⊕			⊕	
Major Concepts																							
scale					○			●		●						→		→					
change					○												●					●	
symbols										●					●	●			●	●			
system													○									○	
cause/effect		○	⊕								⊕		⊕										
time									○				○				⊕	⊕		●	●		
interaction									○				○	○	○		○		○	○	⊕		●
evidence	⊕	⊕	⊕			●	●	●	●	→						→					→		
Geographical Topics/Concepts																							
direction	○		→				→	→							→					→			
land use							○			⊕				→			→					●	
settlement location								⊕														⊕	
services								⊕	●	●		●											
fieldwork						⊕	⊕	●															
distribution		○					⊕	⊕															
urban areas							○		⊕	⊕													
industry					○					○												⊕	
traffic/transport				⊕	⊕	⊕			●													●	
environment/landscape	○				○									○		○	○			○		○	○
erosion/deposition																	⊕			⊕			
ideologies																							
trade				○	○					○	○											○	
weather														○	⊕							→	→
resources										○	○												
regions	⊕					⊕						⊕		⊕	⊕								
irrigation																				○	⊕		
plate movements																			○	○			
leisure/tourism				○	○		○		○	⊕													
natural hazards																		⊕		⊕	⊕	●	
farming						⊕									→	→		→					
government help													⊕									●	
hydro-electricity																						●	○
continent	⊕											⊕			⊕	●							
relief	⊕														⊕		●	●				●	
drainage	○														○								
population		⊕	⊕					●	●	●												⊕	●
migration			⊕										●										
competition/collaboration				○	○					○	○											○	
catchment area					○		○	⊕	●														
central place								○	⊕	●													
GDP											⊕	●	●										

Key   ○ concept / issue / skill introduced or mentioned   ⊕ concept / issue / skill focused on   ● concept / issue / skill / reinforced and taken

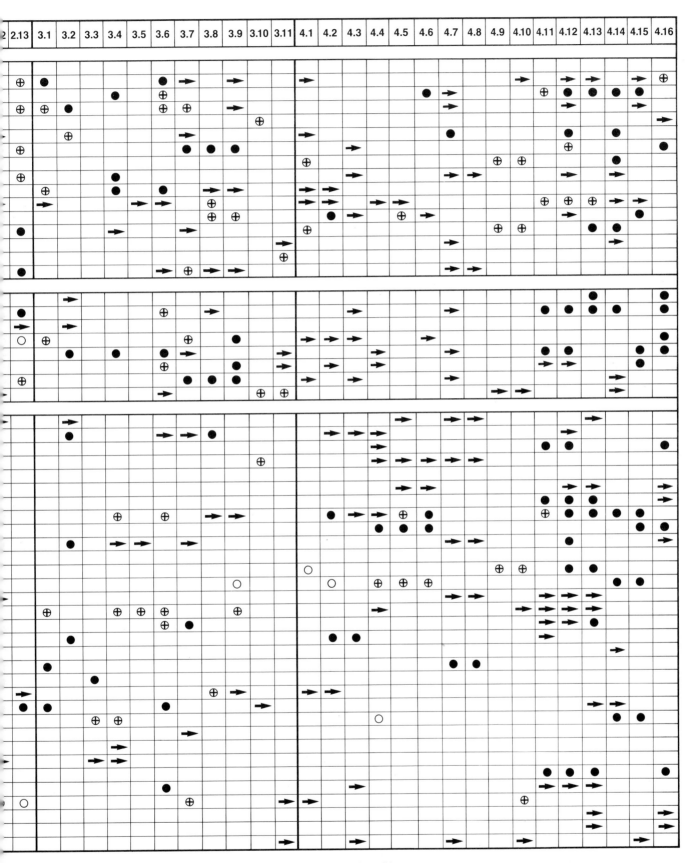

students' knowledge base   ➡ concept / issue / skill expanded and developed into new area

Matrix of Skills

SKILLS \ UNIT NO.	1.1	1.2	1.3	1.4	1.5	1.6	1.7	1.8	1.9	1.10	1.11	1.12	1.13	2.1	2.2	2.3	2.4	2.5	2.6	2.7	2.8	2.9	2.10
Communication and Numeracy																							
advertising/poster design									○									●					
assessment/evaluation		⊕																					
calculating mean														○									
calculating population density		○																					
collecting data		○						⊕															
discussion		○			⊕																		
enquiry/research														○	○								
fieldwork						⊕		⊕															
imaginative writing					⊕				⊕								⊕	⊕					
keyword planning				○																			
measuring the weather																							
observing																							
presenting results						○												⊕					
recording																							
using questionnaire						○		⊕															
testing ideas/hypotheses						⊕																	
interpreting		○												⊕								●	
Map and Graphicacy																							
atlas work	⊕			●		●			●		●	●											
cross-section																				○			
long profile																							
direction																							
drawing graphs — line										○													
drawing graphs — divided bar																							
drawing graphs — bar									●														
drawing graphs — pie									●														
interpreting graphs — line																							
interpreting graphs — bar						○	⊕																
interpreting graphs — pie																							
interpreting graphs — picture																							
interpreting graphs — divided bar																							
map reading/interpretation	○		○				→	→	●	●													
map making and design	○							→															
using satellite information																							
sketching/perspective																							
sketching from photo												⊕											
time line																				⊕			
timetable						⊕																	
reading/using diagrams																				⊕			
using grid references																							
using statistics									○														
using photographs		○								⊕		⊕		⊕									⊕
using scale					○		→	→															
weather chart/symbols																							
Decision Making/Problem Solving																							
deciding on priorities					○																		
grouping/categorising						⊕																	
planning		○																					
predicting/forecasting																			⊕				
recognising problems					○																		
Social																							
group work			○		⊕			→															
cooperating/sharing					○																		
Values and Attitudes																							
awareness of environment					○																	●	●
awareness of others					○																		
empathy			○		○							⊕											
simulation/role play																							

Key ○ concept / issue / skill introduced or mentioned   ⊕ concept / issue / skill focused on   ● concept / issue / skill reinforced and taken as

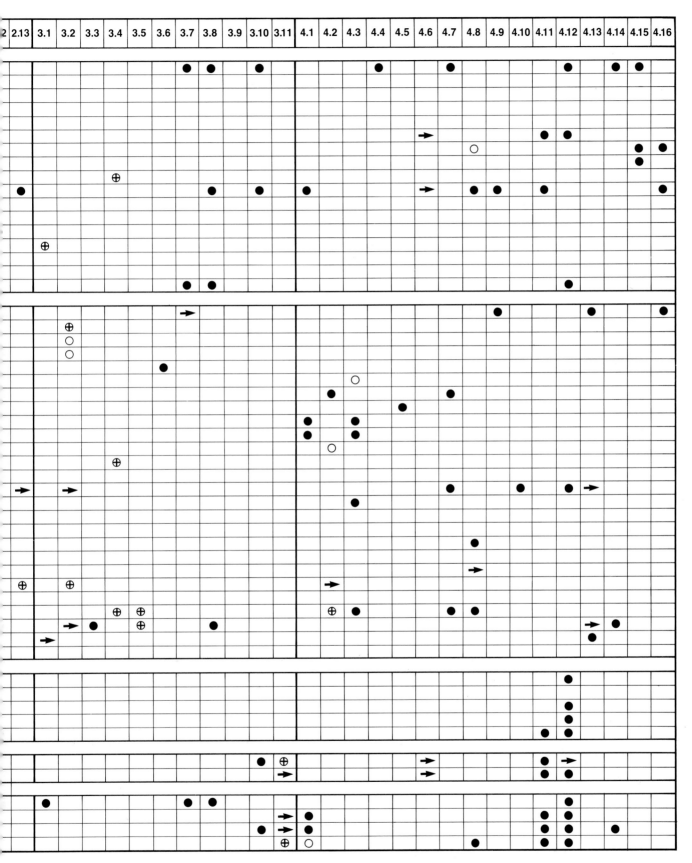

udents' knowledge base ➜ concept / issue / skill expanded and developed into new area

Index of Places